EVOLUTIONARY BIOLOGY OF

Parasites

MONOGRAPHS IN POPULATION BIOLOGY

EDITED BY ROBERT M. MAY

1. The Theory of Island Biogeography, by Robert H. MacArthur and Edward O. Wilson

2. Evolution in Changing Environments: Some Theoretical Explorations, by Richard Levins

3. Adaptive Geometry of Trees, by Henry S. Horn

4. Theoretical Aspects of Population Genetics, by Motoo Kimura and Tomoko Ohta

5. Populations in a Seasonal Environment, by Stephen D. Fretwell

6. Stability and Complexity in Model Ecosystems, by Robert M. May

7. Competition and the Structure of Bird Communities, by Martin Cody

8. Sex and Evolution, by George C. Williams

9. Group Selection in Predator-Prey Communities, by Michael E. Gilpin

10. Geographic Variation, Speciation, and Clines, by John A. Endler

11. Food Webs and Niche Space, by Joel E. Cohen

12. Caste and Ecology in the Social Insects, by George F. Oster and Edward O. Wilson

13. The Dynamics of Arthropod Predator-Prey Systems, by Michael P. Hassell

14. Some Adaptations of Marsh-nesting Blackbirds, by Gordon H. Orians

15. Evolutionary Biology of Parasites, by Peter W. Price

EVOLUTIONARY BIOLOGY OF
Parasites

PETER W. PRICE

PRINCETON, NEW JERSEY

PRINCETON UNIVERSITY PRESS

1980

QL
757
.P74
1980

Preface

"The challenge of diversity" as Mayr (1974) put it, is still with us. Mayr (1976) has noted that the diversity of life is spectacular and its study requires mental exercises not generated in other sciences. Requiring a blend of systematics, ecology, behavior, genetics, and geography, the study of diversity poses one of those "middle-sized questions" that appealed to Bonner (1965) because the whole of biology becomes involved in the answers.

Some of the aspects of diversity such as the better representation of taxa in tropical latitudes have received considerable attention (e.g. Pianka, 1966), and the theory of island biogeography (MacArthur and Wilson, 1967) has contributed enormously to solutions of questions on diversity. Other questions such as the reasons for the varying sizes of taxa have hardly surfaced as debates in biology. Where attempts have been made to answer such questions, they have been qualitative rather than quantitative (Mayr, 1974).

Having been involved with research on a family of parasitic Hymenoptera, thought to be represented by some 60,000 species (Townes, 1969), I have considered questions on the evolution of diversity. No entomologist can escape these questions when such a large proportion of the world's fauna so far described consists of insects. Two events stimulated my interest further. In 1971 Richard Askew published his excellent book on parasitic insects that treated parasitoids and insect parasites on vertebrates for the first time in one volume. Apart from the fund of knowledge and the ecological approach in this book, one thing struck me as important. He mentions that many Hymenoptera in the superfamily Chalcidoidea and some in the superfamily

Cynipoidea were parasitic. Actually, almost all the species in these taxa are parasitic, but some feed on plant tissues (e.g. gall wasps and seed wasps) and others on animal tissues (parasitoids). If parasitic taxa span the plant and animal kingdoms, so must the parasitologist. The second idea came in conversation with Guy Bush, for it became apparent that the tephritid flies he knew so well behaved in ways similar to the cocoon parasitoids I had studied, although the fruit flies utilize plants as hosts and the parasitoids attack animals. The common feature here was utilization of a discrete, small resource such as a fruit, a seed, or a cocoon. When the female is free-living, she can protect her progeny within the food package by leaving a deterrent chemical, thus reducing competition. With such evidence of evolutionary convergence between parasites on plants and animals we agreed that a symposium on parasitic insects would be profitable. This was organized for a national meeting of the Entomological Society of America in 1974 and subsequently published as "Evolutionary Strategies of Parasitic Insects and Mites" (Price, 1975a). The participants, Richard Askew, Guy Bush, Don Force, Daniel Janzen, Robert Matthews, Rodger Mitchell, and Bradleigh Vinson contributed significantly to a synthesis of the evolutionary biology of insects parasitic on plants and animals. It became evident that a large number of insects were parasitic, some on plants, and others on animals and that the diversity of insects and life in general may be at least partly understood by gaining knowledge of the parasitic way of life. The stage was set for a comparative approach to the phenomenon of parasitism.

Parasites are more of a phenomenon than I had thought! They are exceedingly numerous in species and numbers of individuals per species, some taxa having undergone the most spectacular adaptive radiations. New adaptive zones have been frequently created throughout evolutionary time and repeatedly colonized by new parasites with only slight modification from the free-living mode of life. Parasites affect the life and death of practically every other living organism.

How, then, can one small book cover such a diverse array of organisms and interactions between organisms? This book represents an attempt to find the features common to parasites. The task necessitates a synthesis of evolutionary biology and the many disciplines involved with the study of parasites. This synthesis, I believe, leads into several unconventional areas of ecology that deserve more attention. The emerging generalizations represent my attitude toward parasites that, while I find helpful, I do not regard as the only valid point of view. Those who generalize are always faced with those who exceptionalize. Bonner (1965:15) made a plea to the latter: "Yet when we make generalizations about trends among animals and plants . . . it is almost automatic to point out the exceptions and throw out the baby with the bath. This is not a question of fuzzy logic or sloppy thought; it is merely a question whether the rule or the deviations from the rule are of significance in the particular discussion." I do not claim that any principle will apply to all parasites, but rather that it is relevant to more parasites than those to which it does not apply. In addition, discovery of exceptions should be followed by a search for explanations, and these may then help to support the generality or cause its modification to a more useful form. Science proceeds as a progression of approximations toward the truth. Early endeavors, of which I think this is one, may turn out to be far from the truth in many ways, but they are justifiable, even when wrong, if further study is stimulated.

Parasites are so abundant and diverse that examples

could be found to defend practically any thesis. They are also small and relatively specialized and thus share with other organisms of such size and specificity a multitude of problems because of these features, independent of the parasitic mode of life. Such correlated attributes make discovery of the essence of parasitism more difficult than one might think. While I feel that parasites possess a unique set of features, as I explain in Chapter 1, many of the attributes I discuss will be commonly found in other small and specialized organisms, although mention may not be made of this at the time.

Much of this book was prepared during 1977 and 1978 while I was on sabbatical leave from the Department of Entomology at the University of Illinois. I am grateful to the university for supporting my leave. Financial support was also provided by a fellowship generously awarded by the John Simon Guggenheim Foundation. This facilitated a seven-month visit to the British Isles and continental Europe and a shorter time in Central America, where I gained enormously by meeting many scientists who shared their knowledge with me. I am most grateful for the opportunities and support the fellowship provided. The monograph was completed with support from a National Science Foundation grant DEB 78-16152.

Many colleagues have contributed significantly to this book. I am grateful to those who read earlier drafts in whole or in part and provided critical comments and positive response: Roy C. Anderson, Roy M. Anderson, Paulette Bierzchudek, Carl Bouton, Paul Gross, John Holmes, Peter Kareiva, John Lawton, Robert May, Bruce McPheron, Beverly Rathcke, Richard Root, John Thompson, Jeffrey Waage, and Arthur Weis. These and others were generous enough to provide information, important leads, reprints, and preprints of papers: Richard Askew, Jeremy Burdin,

Guy Bush, Howard Cornell, Doug Futuyma, John Harper, John Jaenike, Clive Kennedy, Robert Morris, Patrice Morrow, Klaus Rohde, John Schneider, Don Strong, A. C. Triantaphyllou, John Whitaker, Tom Whitham, and Peter Yeo. For the artwork I am grateful to Alice Prickett, Mike Paulson, and Carol Kubitz. Donna Mohr, Peggy Vaughan, and Donna Stowe all helped with typing and other essentials for which I am most thankful. For supervising a department in which the climate is so conducive to enterprises such as this, I thank Stanley Friedman. To Maureen, Gavin, and Robin goes my deep appreciation for creating a home environment in which this monograph could be written.

Contents

Preface   v

1. Introduction: The Parasite's Lot in
   Evolutionary Biology   3

2. General Concepts   15

3. Non-Equilibrium Populations and Communities   44

4. Genetic Systems   76

5. Adaptive Radiation and Specificity   105

6. Ecological Niches, Species Packing, and
   Community Organization   134

7. Parasite Impact on the Evolutionary Biology
   of Hosts   149

8. Further Study   171

   Bibliography   177

   Author Index   227

   Subject Index   233

EVOLUTIONARY BIOLOGY OF

Parasites

CHAPTER ONE

Introduction: The Parasite's Lot in Evolutionary Biology

Parasites form a large proportion of the diversity of life on the earth. Therefore in attempts to reach generalizations or to identify patterns in biology they should have received a preponderance of attention. This has not been the case largely because small organisms in general are poorly known, but also because many biologists do not appreciate the commonness of the parasitic way of life. Also, we tend to see organisms as if they lived within the limits of our own limited experience. But of course parasites live in a very different world from ours, more like the worlds of the water-beetle and the bacillus described by Thompson (1942). As he says "we have come to the edge of a world of which we have no experience, and where all our preconceptions must be recast" (1942:77).

Indeed, I think there is much room for recasting the image of parasites held among biologists, and in so doing certain views in ecology, evolutionary theory, and parasitology must also be recast. It has not been generally realized that the most extraordinary adaptive radiations on the earth have been among parasitic organisms. Ecological and evolutionary principles should address such spectacular speciation and provide the basis for an understanding of the mechanisms involved. I do not think that a coherent body of theory exists for the evolutionary biology of parasitic organisms.

This recasting cannot be evaluated fairly without a re-

3

view and a reappraisal of the conventional wisdom on parasites as a basis for understanding the evolutionary biology of parasites. This chapter is largely devoted to such tasks, concentrating on three areas: the definition of a parasite, parasite ecology, and parasite evolution.

There are probably as many definitions of a parasite as there are books on parasitism. Rather than construct my own definition and incorporate my own biases or preconceptions I resort to *Webster's Third International Dictionary* for what must be a generally acceptable definition: *a parasite is an organism living in or on another living organism, obtaining from it part or all of its organic nutriment, commonly exhibiting some degree of adaptive structural modification, and causing some degree of real damage to its host.* The *Oxford English Dictionary* definition is similar. It must be emphasized that an individual of any parasitic species will usually gain the majority of its food from a single living organism. Although a species of parasite may utilize several or many host species, each individual obtains most of its nutrition from an individual host. Species with complex life cycles may exploit two or three hosts in a predictable sequence. This contrasts with the more generalized grazers, browsers, and predators that feed on many organisms during their life time and with saprophages that feed on dead organic matter.

Anderson and May (1978) argue that an organism should be classified as a parasite only if it has a detrimental effect on the intrinsic growth rate of its host population. This is a realistic operational definition for those concerned with modelling ecological impact of parasites on host populations. But my focus is on the parasite itself and its evolutionary biology in which intimate association with and unfavorable impact on its host, however small, are the crucial qualities of its life style.

4

When the above definition of a parasite is applied objectively without taxonomic or disciplinary constraints, many organisms must be recognized as parasites. Many insects that feed in or on plants fit the definition well. The large order Homoptera including leafhoppers, froghoppers, aphids, scale insects, and whiteflies is composed almost completely of parasitic species. The larvae of the even larger order Lepidoptera usually feed and mature on a single individual of the host plant species and gain a large percentage of the total nutritional requirements for the organism's life span. Other orders such as the Thysanoptera, Hemiptera, Coleoptera, and Diptera swell the ranks of parasitic insects on plants, to which should be added the large numbers of mites, nematodes, fungi, bacteria, and viruses of interest to the plant breeder (e.g. Day, 1974). Parasitic angiosperms are frequently ignored when parasitism is discussed (Kuijt, 1969).

Parasites on animals include those of interest to the conventional parasitologist: viruses, bacteria, protozoa, flatworms (flukes and tapeworms), thorny-headed worms (Acanthocephala), nematodes, and arthropods (crustaceans, insects, mites). Other parasites such as the parasitic Hymenoptera and Diptera are of more interest to the entomologist and specialist in biological control.

A few organisms that have been regarded as parasites should also be excluded. Blood-sucking organisms are often considered as parasites, but mosquitoes and black flies have very brief contact with the animal they feed on so that the relationship should not be regarded as parasitic. Other blood feeders remain on an animal for a considerable time such as many fleas (e.g. see Rothschild and Ford, 1973), ticks (e.g. Arthur, 1965; Hoogstraal, 1967), and sucking lice and may be considered as parasites. However, as pointed out by Askew (1971) and Kennedy (1975a), there are no discrete limits to the parasitic habit that are biologically meaningful. Ticks utilize several hosts in the course of a lifetime. The pollinating fig wasps are parasitic in their habits except that benefits to the host plant outweigh the damage, making them, in the balance, mutualists. Some fungi, such as *Rhizoctonia solani* and *Armillaria mellea* are important plant parasites, each attacking at least 100 plant species (see Moore, 1959), but they also become mycorrhizal on orchids (Harley, 1969). Indeed Garrett (1970) stresses that initially the seedling orchid must be parasitic upon the mycorrhizal fungus. Thus the majority of this book will deal with organisms that fit the definition quite well, but on occasion those more loosely associated with hosts will be considered. Social parasites and brood parasites represent phenomena of a different kind and will remain largely unconsidered.

Thus parasite species are incredibly numerous, and they probably affect every living organism at some stage in its lifetime. Yet it is very difficult to obtain a clear picture of how many parasite species there are compared to those that are free-living. Parasites are small and usually cryptic and unobstrusive members of any biota. Probably the most concerted regional effort on a taxon represented by both parasitic and free-living species has been on insects in the British Isles. A check list of British insects by Kloet and Hincks (1945) permits an evaluation of the commonness of the parasitic habit (Table 1.1). Of the 20,244 insect species listed in the check list, 16,929 can be readily classified into the categories in the table. For the remainder there is a diversity of feeding habits within a family or poor information on food eaten by immature and adult stages, so they have not been classified.

The categories in Table 1.1 are designed to cover all possible feeding types including the self-explanatory head-

INTRODUCTION

TABLE 1.1 Feeding habits of British insects based on analysis of the check list of British insects by Kloet and Hincks (1945) (After Price, 1977).

Order	Predators	Nonparasitic herbivores and carnivores	Parasites		Sapro-phages
			On plants	On animals	
Thysanura					23
Protura					17
Collembola					261
Orthoptera		39			
Psocoptera		70			
Phthiraptera				308	
Odonata	42				
Thysanoptera			183		
Hemiptera	123		283	5	
Homoptera			976		
Megaloptera	4				
Neuroptera	54				
Mecoptera	3				
Lepidoptera			2,233		
Coleoptera	215	65	909	18	1,637
Hymenoptera	170	241	435	5,342	36
Diptera	54	231	922	311	1,672
Siphonaptera				47	
TOTALS	665	646*	5,941	6,031*	3,646
% OF INSECT FAUNA	3.9	3.8*	35.1	35.6*	21.5

* These figures differ slightly from those in Price (1977) because the biting flies (Diptera) have been transferred to the category of Nonparasitic herbivores and carnivores.

ings of predators and parasites. The nonparasitic herbivores and carnivores include browsing insects (e.g. grasshoppers, Orthoptera, and bark lice, Psocoptera), pollinators (e.g. bees, Hymenoptera), and insects that quickly take a blood meal from the host (e.g. mosquitoes and blackflies, Diptera). Saprophages include the genuine feeders on dead organic matter (e.g. dung and carrion beetles, Coleoptera) and those species that feed on microorganisms in decaying organic matter (e.g. fruit flies, Drosophilidae, Diptera).

7

INTRODUCTION

The majority of insects are classified in the parasitic groups representing 70.7 percent of the fauna (Table 1.1). This contrasts with Askew's (1971) careful estimate that 15 percent of all insects are parasitic on animals. However, Askew used the described species of the world to obtain his estimate whereas parasites are still very poorly known in many regions. For example, Townes (1969) estimates that only 10 percent of the Ichneumonidae (parasitic Hymenoptera) in the Neotropics are described and even in the Nearctic only about 35 percent may have been described. Thus the discrepancy derives from an inadequate sampling of the world fauna by taxonomists rather than a dichotomy in our views on which insects are parasitic on animals.

Even if a conservative estimate is made by assuming that all the parasites in the fauna have been identified in my classification, they represent about 60 percent of insects. Since about three-quarters of the known animals on earth are insects, an estimate that parasitic insects represent close to half the animals on earth does not seem unrealistic. When other large groups of parasitic animals found among the nematodes, flatworms, mites, and protozoa are added to the number of parasitic insects, it is clear that parasitism as a way of life is more common than all other feeding strategies combined. Indeed, with the world sampling of nematodes, mites, protozoa, and other groups with many parasitic species in such a juvenile state we should expect the proportion of known parasitic species to increase and the supremacy of insects as the most diverse taxon of animals to be severely challenged. As May (1978b) illustrates, while acknowledging the inadequacy in sampling of small species, there is a general relationship between number of species extant (S) and the inverse square of their typical linear dimension $(S \sim L^{-2})$.

Because of their small size and specialist habits, the ecology of parasites differs considerably from completely free-living forms, as I emphasize in Chapter 2. But Elton (1927:75) in his pioneering work on animal ecology concluded that "it is best to treat parasites as being essentially the same as carnivores [i.e. predators]. . . . The resemblances between the two classes of animals are more important than the differences." This attitude is now prevalent in English ecology texts (e.g. Andrewartha and Birch, 1954; Odum, 1971; Krebs, 1972; Colinvaux, 1973), and although much attention is devoted to predation, parasitism is almost ignored (see also Anderson, 1979). Other ecologists (e.g. Naumov, 1972; Kennedy, 1975a) recognize parasitism as a discrete mode of life but have not addressed the opposing school. But the study of parasites carries us into areas of ecology where students of predators and other generalists have rarely led. Such areas remain relatively neglected subjects in ecology and include: (1) non-equilibrium breeding populations and communities (see Chapter 3); (2) genetic systems as adaptive syndromes of species (see Chapter 4); (3) the ecology of symbiosis (see Chapters 5 and 7); (4) the role of one species in modifying the antagonistic interaction (competition, predation) between two other species (see Chapter 7).

Probably the most important feature of parasite ecology is the close association with negative impact an individual maintains with another living organism. As a result the host in many ways acts as the environment for the parasite, as repeatedly pointed out by parasitologists and others (e.g. Noble and Noble, 1976; Salt, 1961), but it is an environment that reacts defensively and individually over the short term to the impact of the parasite (e.g. immune responses) and reacts over the long term as a population and species coevolving with its parasite population. With

9

such intimate suites of action, reaction and counteraction coevolution will be more precise than in any pollinator-plant relationship (except perhaps those of fig wasps or yucca moths) and will probably necessitate a greater genetic commitment to coevolution on both sides than many mutualistic relationships. Short generation times and large populations relative to hosts permit a very precise tracking of the hosts' genetic control of resistance with escalation in attack proceeding rapidly after the evolution of new defenses. An exceedingly dynamic coevolutionary system results (e.g. as summarized for the myxoma virus of rabbits by Fenner and Ratcliffe, 1965).

Several important ecological attributes of parasites stem from this close relationship with the host. As a resource available for exploitation by parasites the range of hosts utilized is often relatively well-known compared to the range of a food resource utilized by free-living organisms such as birds or small mammals. Thus measures of specialization and generalization, niche breadth and overlap, species packing and the community matrix can be obtained by knowing relationships between hosts and parasites. The tight coevolutionary systems frequently lead to high specificity among parasites so uncharacteristic of free-living organisms. Switching of food, modifiable functional response, and movement to high host densities are strategies frequently not available to parasites in ecological time, but frequently invoked by ecologists as mechanisms maintaining population and community stability. While the parasite is associated with the host, many aspects of the host's ecology will change since the parasite frequently alters the host quality. Behavior, reproductive potential, competitive ability, and susceptibility to predation may all be modified and the degrees of change will differ with levels of infection. This variability in the host population, superimposed on the genotypic and phenotypic differences within it, greatly complicates its ecological and evolutionary processes.

The commonest view on the evolution of parasites is that they evolve slowly and represent dead ends in any phylogeny. In the last principle in their parasitology text—the final thought left with the reader—Noble and Noble (1976:525) state that "parasites as a whole are worthy examples of the inexorable march of evolution into blind alleys." Mayr (1963:596) echoes the same sentiment when he states that "most specialization leads into blind alleys," although he adds that some highly specialized types have evolved to exploit new adaptive zones. Huffaker (1964:645) also believes that "specialization is a deepening rut in evolution." There are probably at least four lines of reasoning leading to this view.

One so-called evolutionary principle is that more generalized organisms and phylogenies (clades) persist through time for periods longer than specialized organisms and phylogenies. Cope (1896) proposed his law of the unspecialized where evolutionary advance is made from unspecialized forms and specialization leads to dead ends. Thus the conclusion is inevitable that extinction rates in parasitic taxa will be much higher than in predatory taxa. Although Simpson (1953) has challenged the principle saying that the study of fossils has demonstrated the extinction of both specialized and generalized clades, the appealing logic has attracted several authors (e.g. Newell, 1967; Cifelli, 1969; Bretsky and Lorenz, 1970, Valentine, 1973) as pointed out by Flessa, Powers, and Cisne (1975). Testing more rigorously than previously the questions of whether generalists survive longer than specialists and if the degree of specialization is the focus for the selective processes that induce extinction, Flessa, Powers, and Cisne (1975) concluded that a negative answer was justified for both. Simpson's view has been upheld and the principle should be laid to rest in the minds of the palaeobiologist and the parasitologist.

Another apparent trend in evolutionary time is a general increase in size of organisms, the early recognition of which may again be ascribed to Cope (1885, 1896). Newell (1949) added support to the principle and found no inverse trends among invertebrate fossils. With increased size go increased complexity and greater control over the environment, attributes associated with biological progress (Huxley, 1942, 1953). Again, parasites run counter to the upward trend in size and complexity and thus have been cast as evolutionarily retrogressive. But trends seen in morphological attributes are inadequate for estimating evolutionary potential as we shall see in the next paragraph and in Chapter 2.

Rates of evolution until recently were largely measured in terms of morphological change (see Simpson, 1953). Those taxa with considerable morphological complexity were found to be evolving more rapidly than simpler organisms. Thus parasites, which are small with frequently highly simplified morphology and relatively few characters on which to base a taxonomy, by association with this trend must evolve slowly, if at all. However, Schopf et al. (1975) have demonstrated convincingly that morphological complexity is simply correlated with apparent rates of evolution, a correlation stemming from the methodology rather than from real biological phenomena (see also Raup et al., 1973). Morphological change is an inadequate measure of genetic change and therefore a poor estimator of evolutionary change. This is so particularly for parasites where host topography dictates morphology to a large extent, pressures for conformity to a particular pattern are enormous, and we see many examples of convergent evolution such as among ectoparasites of mammals and birds.

Finally, the concept of slow evolution in parasites may result in part from the knowledge that asexual reproduction and inbreeding are commonly found and lead to reduced variability in a population with consequent slower changes in the gene pool. Although there is some evidence of relatively low genetic diversity in parthenogenetic organisms (e.g. Pamilo, Vespäläinen, and Rosengren, 1975; Snyder, 1974; Crozier, 1970; Oliver, 1971), there appear to be ample variability on which natural selection can act and quite surprising amounts of variation in polyploid species (e.g. Suomalainen, Saura, and Lokki, 1976). This contrasts with what population genetics theory predicts and should cause scepticism about conclusions on rates of evolution in asexually reproducing organisms. In addition many factors in the genetic systems of parasites discussed in Chapter 4 compensate for the potential for reduced variability through asexuality or inbreeding.

Many biologists tend to think anthropocentrically, along with Huxley (1942, 1953), of biological progress as the increasing control of the environment and independence from it. Thus the mammals and birds, and particularly man, are the glories of the evolutionary process. Parasites have been viewed in contrast as if they were in blind alleys stemming from an otherwise progressive evolutionary line. But is an alley that results in the evolution of 60,000 species in a single family blind? Isn't a mode of exploitation that leads to the evolution of more than half of all animals on this earth more "progressive," and in the end more viable than one that has led to a mere 12,700 species of birds and mammals? Huxley (1953:127) himself considered progress as "constantly leading life into regions of new evolutionary opportunity." The opportunities for the great apes, includ-

13

ing man, are small compared to the immense adaptive radiations still underway among parasitic organisms (Chapter 5).

The fault in Huxley's views is that no biological progress exists in the evolutionary process. Darwin's theory of natural selection "explodes any concept of inherent progress" (Gould, 1975:826; see also Gould, 1974). As White (1973:760) has stated, the process results in a series of adaptive reactions to ecological opportunities, so the evaluation of evolutionary potential is perhaps a more valid way of comparing exploitative and reproductive strategies. I shall attempt to establish in this book that no group of organisms on this earth can surpass the parasites in their potential for continued adaptive radiation.

CHAPTER TWO

General Concepts

Darwin (1872: Chapter 4) emphasized the relationships between organisms as being most influential in the generation of organic diversity as follows: "As the number of species in any country goes on increasing, the organic conditions of life must become more and more complex. Consequently, there seems at first sight no limit to the amount of profitable diversification of structure, and therefore, no limit to the number of species which might be produced." His emphasis is adopted in this book since Darwin's view seems particularly relevant to the evolution of parasites. Darwin did recognize however an upper limit to diversification as discussed later in this chapter.

Darwin's statement also accentuates the need for understanding the environment where species evolve. Thus for an evolutionary perspective of parasites we must have an ecological basis from which to work. Using Hutchinson's (1965) metaphor, we view the evolutionary play as staged in an ecological theater. The six general concepts proposed here are therefore divided into three that attempt to depict the typical ecological setting in which parasites live and three that summarize the probable evolutionary consequences of species living under these conditions.

But the ecological setting must be perceived as the parasite experiences them and not necessarily in the conventional ways comfortable to an ecologist. For example, most ecologists emphasize the factors leading to stability and an equilibrium state, being justifiably interested in total numbers of two interacting species spread over many environ-

15

mental patches, just as in Huffaker's (1958) orange mite and predatory mite experiments. By contrast the effective environment for a parasite is the patch where it is situated and another patch that it or its progeny must reach in order to find new hosts. Thus our ecology must become one of within-patch dynamics and between-patch dynamics rather than an ecological overview of the many patches so successfully employed by analysts of predator and prey interactions and ecosystems models (e.g. Hassell, 1976, 1978; May, 1973).

The following basic concepts serve as a beginning for the synthesis of parasitology and evolutionary biology (using the word *parasitology* to mean the study of all parasites). The general approach compares the ecology and evolution of parasites with that of predators. All but Ecological Concept 3 have been proposed before (Price, 1977). They are necessarily simple so that generality is retained, but they encapsulate what I think is the essence of the parasitic way of life. They portray the mechanisms that have led to the amazing diversity of parasitic organisms.

ECOLOGICAL CONCEPTS

Concept 1. *Parasites are adapted to exploit small, discontinuous environments.* For a parasite each host exists in the matrix of an inhospitable environment (Williams, 1975). Each host population is also relatively distant from others. For very small organisms a wide dispersion of resources within patches and considerable distances between patches makes colonization of new hosts hazardous. Adaptations to reduce this hazard include: (1) mass production of spores or eggs (e.g. rusts and tapeworms, see also Jennings and Calow, 1975); (2) dispersal of inseminated females that form a high proportion of the population (e.g. tetranychid

16

mites, Mitchell, 1970); (3) dispersal by attaching to a larger organism (phoresy), often adults of the host species, that makes host discovery accurate (e.g. tarsonemid mites, Lindquist, 1969; parasitic insects, Clausen 1976). In each of these cases a single female can found a colony on a new resource remote from its origin. Multiplication in the colonizer's progeny may lead to a new and relatively isolated population. New propagules colonize other isolated resources. Thus parasites tend to exist in small homogeneous populations with little gene flow between them (see also Jones, 1967).

Dispersal in time is also achieved among parasites by the extreme longevity of resting stages, a strategy shared with many patchily distributed colonizing species of plants (that act as hosts to the parasitic fungi and nematodes given here as examples). Sclerotia of the parasitic root fungus *Phymatotrichum omnivorum* show little loss of viability after 12 years (Garrett, 1970). Chlamydospores of *Fusarium oxysporum* remained infective after 20 years (Rishbeth, 1955). Cooper and Van Gundy (1971) indicate that plant parasitic nematodes may remain quiescent for 23 years and in a cryptobiotic state for 39 years. Seeds of parasitic flowering plants remain viable for decades (Sunderland, 1960). Among parasites of animals longevity can be as extreme. Some pathogenic viruses and bacteria remain viable for decades. Resting spores of the fungus *Massospora cicadina* must be infective for at least 17 years in order to bridge the gap between availability of their periodical cicada hosts (Lloyd and Dybas, 1966). A cecidomyiid *Sitodiplosis mosellana* may remain dormant for 13 years, and some of its parasitoids for 6 to 8 years (Barnes, 1958). Metacercariae of the digenean trematode *Diplostomum spathaceum* are known to survive for up to 5 years (Kennedy, 1976).

Types of reproduction in parasites that permit a single

17

female to found a new population include inbreeding among progeny (e.g. parasitic wasps, Askew, 1968), hermaphroditism (e.g. nematodes, trematodes, cestodes) or asexual reproduction (e.g. polyembryony in digenetic trematodes and parasitic wasps, parthenogenesis in many taxa). Parthenogenesis is very common among parasites (see also Ghiselin, 1974). For example, in the parasitic arthropods arrhenotoky (the production of males from unfertilized eggs) occurs in all Hymenoptera, probably the majority of Thysanoptera, many iceryine and aleurodid Homoptera, and in some Coleoptera (White, 1973). It is a major mode of reproduction in certain families of mites (Oliver, 1971). Thelytoky (the production of females from unfertilized eggs) occurs sporadically in a great variety of insects (Suomalainen, 1962), and cyclical parthenogenesis (or heterogonic life cycles, Williams, 1975) is prevalent in aphids (Aphididae) and gall wasps (Cynipidae), and it occurs in gall flies (Cecidomyiidae) (Suomalainen, 1962; White, 1973) and in many parasites in other taxa (Williams, 1975). Thus parthenogenesis occurs in five of the largest families of parasitic insects on plants in Britain (Table 2.1): Cecidomyiidae, Curculionidae, Aphididae, Tenthredinidae and Cynipidae (see Suomalainen, 1962). In the animal parasites of the British fauna eight of the largest families have all species parthenogenetic since they are within the order Hymenoptera: Ichneumonidae, Braconidae, Pteromalidae, Eulophidae, Platygasteridae, Encyrtidae, Diapriidae, and Ceraphronidae (Table 2.1).

Parthenogenesis is rarely found in predatory groups such as Odonata, spiders, and predatory Heteroptera. Of the ten largest families of predators in the British insect fauna only two hymenopterous families, Sphecidae and Vespidae, show any form of parthenogenesis (cf. Suomalainen, 1962; White, 1973).

18

TABLE 2.1 Number of species in the ten largest families in the British insect fauna in the categories of predators, herbivorous parasites, carnivorous parasites (from check list by Kloet and Hincks, 1945). Families marked with an asterisk contain some or all members reproducing parthenogenetically. The family Cynipidae contains some carnivorous parasites (after Price, 1977).

Predators		Herbivorous Parasites		Carnivorous Parasites	
Dytiscidae	110	Cecidomyiidae*	629	Ichneumonidae*	1938
Sphecidae*	104	Curculionidae*	509	Braconidae*	891
Coccinellidae	45	Aphididae*	365	Pteromalidae*	649
Corixidae	32	Tenthredinidae*	358	Eulophidae*	485
Cucujidae	32	Noctuidae	298	Tachinidae	228
Hemerobiidae	29	Chrysomelidae	248	Philopteridae	176
Vespidae*	27	Cicadellidae	242	Platygasteridae*	147
Asilidae	26	Cynipidae*	238	Encyrtidae*	144
Anthocoridae	25	Olethreutidae	216	Diapriidae*	125
Saldidae	20	Miridae	186	Ceraphronidae*	108
MEANS	45		329		489

The suite of adaptations for coping with a very patchy resource will be treated in greater detail under genetic systems of parasites in Chapter 4. Patchy resources are not unique to parasites. The problems posed by such resources are faced by other small organisms exploiting such things as dung, carrion, and rotting logs (c.f. Hamilton, 1978).

Concept 2. Parasites represent the extreme in specialized resource exploitation. The small size of parasites, the specific cues used in discovery of relatively large food items (e.g. see Vinson, 1975; Galun, 1975; MacInnis, 1976; Holmes, 1976), and the poor mobility of many stages in the life cycle mean that they must exploit a coarse-grained environment in the sense of MacArthur and Levins (1964) and MacArthur and Wilson (1967). Such small organisms facing relatively large habitat differences compared to the tolerance of individuals may respond in two ways to environmental variability (Levins, 1962). If the environment is stable in time and variable in space, the local population should be monomorphic and specialized with a geographic pattern of discrete races. If the environment is uniform in space and variable in time, a population should be polymorphic, with specialized types, and have a geographic pattern of clines in frequencies of these specialized types (see also Clarke, 1976). Since hosts probably represent a variable resource both in time and space, the development of both geographic races and polymorphism in parasites could be predicted. Both may occur in the same species as illustrated by the extensive and detailed studies by Whitlock and his associates on the nematode *Haemonchus contortus* that is parasitic in the abomasum of sheep (Das and Whitlock, 1960; Crofton, Whitlock, and Glazier, 1965; Glazier, Crofton, and Whitlock, 1967; LeJambre and Whitlock, 1968, 1976; LeJambre et al., 1970, 1972; LeJambre and Ratcliffe, 1971). Specialization permits a relatively large number of species to pack into a given set of resources so many species may coexist in a community.

In contrast to parasites, predators are frequently relatively large and mobile and exploit a relatively fine-grained environment. Habitat differences are small compared to the tolerance of the animals. Levins (1962) predicted that such populations will be monomorphic and unspecialized, with geographic patterns showing continuous clines. Few of these generalists can coexist under equilibrium conditions.

The difference in grain between environments for parasites and predators and the degree of specialization in each feeding type may be seen in their feeding habits. Many species of parasite utilize only one species of host (Table 2.2), and only a small proportion (less than 6 percent) of species in the families given as examples are known to utilize more than nine hosts. Among predators and grazers it is unusual to find a species that utilizes less than ten spe-

TABLE 2.2 Comparison between number of host species utilized by parasite species (given as the percentage of parasite species in each host number class) and the number of species in the diet of predators and grazers (given as the percentage of species in each number class).

Number of hosts	Parasites on Plants — Agromyzidae (1)*	Miridae (2)	Gelechiidae (3)	Parasites on Insects — Ichneumonidae (4)	Braconidae (5)	Parasites on Fish — Monogenea (6)	Digenea (6)	Parasites on Mammals and Birds — Nematoda (7)	Philopteridae (8)	Hippoboscidae (9)	Number of species or larger taxa in diet+	Predators on Seeds — Birds (10)	Predators on Insects — Small mammals (10)	Bats (11)	Grazers — Small mammals (10)	Ungulates (10)
1	57	22	53	53	60	85	25	52	87	17	1					
2	20	14	26+	18	14	3	21	20	9	9	2					10
3	7	13	6+	11	10	3	14	8	2	15	3		9		5	
4	6	21	3+	4	7		11	5	1	13	4		9			
5	4	8		5	2	3	11	1	1	7	5					
6	3	6		3	1	3		1			4	6	3			5
7	<1	7		1	2		7	1			4	7	9			
8	<1	1		1	<1	3		1			4	8	3			14
9	<1	4		1	1		4	<1			2	9	4	17			10
10-19	2	3		3	2		4	5			4	10-19	43	35	25	40	20
20-29	<1	1		<1			4	3			7	20-29	20	26	25	31	10
30-39				<1				1			2	30-39	10		50		10
40-49												40-49	4	4			10
50-59					<1			<1			2	50-59	1				10
60-69								<1				60-69	3				
70-79												70-79				5	
80-89					<1						7	80-89					20
Number of Species	267	72	112	514	214	34	28	340	122	46			70	23	4	22	10

* Sources: (1) Spencer, 1972; (2) Southwood and Leston, 1959; (3) Ford, 1949; (4) Townes and Townes, 1951; and Heinrich, 1960-62; (5) Griffiths, 1964-1968; (6) Kennedy, 1974; (7) Cram, 1927; (8) Theodor and Costa, 1967; (9) Bequaert, 1956; (10) Martin, Zim, and Nelson, 1951; (11) Whitaker, 1972.
+ Frequently only larger taxa are given: for example, the families or orders of insects eaten by bats.

cies as food and most species are situated well down in Table 2.2.

Concept 3. Parasites exist in non-equilibrium conditions. The parasite's world is a small patch of resources surrounded by other small patches where the probability of colonization of any one patch by a parasite or its progeny is very low. In any one patch the resources, being living organisms, are ephemeral, so tenure on a patch is brief. Colonization of new patches carries large risks with potent selection for colonizing ability. The strategic necessities of colonizing species are summarized by MacArthur and Wilson (1967) and treated in detail by Baker and Stebbins (1965).

The ephemeral nature of resources in any one patch is amplified by each additional species essential to the parasite's life system. Each species may remain in a patch a short time but if unsynchronized with other species, the establishment by a parasite is impossible. The frequently complex nature of parasite life cycles, involving several species directly, as well as many species indirectly, means that several species and other environmental conditions must overlap in time and space before any resource becomes exploitable and a patch is created in which a parasite may establish itself. Departure of any one element in the patch renders the parasite population inviable.

For examples of this scenario consider the complex life cycles of schistosomatid digenean trematodes and other heteroecious parasites among aphids and adelgids (e.g. *Pemphigus, Eriosoma, Adelges, Pineus*), fungi (e.g. *Cronartium ribicola, Puccinia graminis, P. hordei*), nematodes (*Splendidofilaria, Wuchereria, Brugia, Onchocerca*), and protozoa *(Plasmodium, Trypanosoma, Leishmania)*. Fasciolid and troglotrematid digenean trematodes utilize three hosts to complete the life cycle. The external stages of the

22

parasite and all the hosts have very different requirements for survival and all must overlap before a patch can be exploited by the parasite. Traub and Wisseman (1974) speak of the "zoonotic tetrad" involved with the epidemiology of chigger-borne rickettsiosis or scrub typhus in tropical and temperate latitudes: (1) the pathogen *Rickettsia tsutsugamushi*, (2) trombiculid mites (chiggers) in the *Leptotrombidium deliense*-group, vectors of *Rickettsia*, and themselves parasitic upon (3) small mammals, especially rats (*Rattus*) that together are found in (4) a stage in ecological succession of vegetation characterized by vegetation on recently disturbed land with scrubby conditions that become rapidly overgrown in tropical plant succession. Additional factors are involved: (5) humans must pass through patches with all four factors present in order to become infected from a mite taking a blood meal and (6) moist conditions, patchy in space and time, favor infection of humans. The occurrence of all factors together is likely to be very local in space and brief in time and highly unpredictable for any of the organisms concerned, including man. This is emphasized by Traub and Wisseman (1974:244): "The characteristic localization of chigger-borne typhus is of a dual nature, comprising endemic districts of a large size (ranging from a few square km to extensive geographic areas, such as islands or valleys) and, within such locales, a markedly patchy distribution of highly circumscribed foci or even microfoci, of typhus islands which may only be a few meters in diameter." Regarding the time factor, these authors point out that an area where no infection was known may become infective within a few months and the infective status of a patch may change dramatically from year to year. Stability within the patch has a remote probability.

We must regard community organization of parasites in a

similar manner to that adopted for parasite populations. Communities within patches, and colonization of patches, deserve most attention. In addition, community equilibrium as defined in MacArthur and Wilson's (1967) island biogeographic model is not a phenomenon experienced by individuals of parasitic species for, as MacArthur and Wilson (1967) and Simberloff and Wilson (1969, 1970) show, tenure of a site by any one species is brief. Not only is there a low probability of colonizing a patch, but a high probability of extinction. Thus, when we consider the ecology of an individual and its progeny, to which their exploitation strategies are adapted, the non-equilibrial state of patches must be emphasized for parasite populations and parasite communities; this is an approach to be expanded upon in Chapter 3.

EVOLUTIONARY CONCEPTS

Concept 1. Evolutionary rates and speciation rates can be high. The environmental constraints on the parasitic habit portrayed in the three ecological concepts promote the fractionation of gene pools and produce an advantage to inbreeding or asexual reproduction. As MacArthur and Wilson (1967:78) stress for colonizing species, "cohesiveness of the propagule is essential." Such constraints foster rapid divergence of populations, race formation, and eventual speciation. Short generation times and high fecundity result in high reproductive rates that permit dramatic changes in population size and rapid differentiation of populations under dissimilar selective regimes. Predominantly homozygous populations (or haploid males in arrhenotokous species) have all genes exposed to selection in each generation. Both the founder effect and genetic drift are prob-

24

ably significant in speciation of parasites. Wright (1940, 1943, 1949) and Carson (1968, 1975) have explored the evolutionary potential of populations with this structure and both have concluded that the probability of evolution and speciation is much higher than in a randomly breeding population of the same size. Powell (1978) provided experimental support for this concept. Other authors have recognized the importance of such situations in the extensive speciation of parasitic wasps (Askew, 1968; Gordh, 1975) and bark and ambrosia beetles (Beaver, 1977).

Evidence for high evolutionary rates may be seen in the many sibling species, subspecies, and host races observed in parasitic organisms (e.g. Thorpe, 1930; Mayr, Linsley, and Usinger, 1953; Brown, 1959). Mayr (1963) states that sibling species are especially common among insects and singles out the Lepidoptera, Diptera, Coleoptera, and Orthoptera as providing abundant examples. In the Coleoptera he notes that sibling species are particularly common in the Chrysomelidae, Cerambycidae, and Curculionidae, three of the largest families in the order, with the majority of species parasitic on plants. Sibling species are very common in the genus *Erythroneura* (Homoptera: Cicadellidae) in which about 150 host transfers have resulted in some 500 species in the genus (Ross 1962). Zimmerman (1960) suggests that five or more species of *Hedylepta* (Lepidoptera) must have evolved within 1000 years in Hawaii since they are endemic and specific to banana which was only introduced that long ago. Bush (1975a) stated that a new race of the western cherry fruit fly *Rhagoletis indifferens* was extant on domestic cherry within 89 years of the plant's introduction to northwest North America. Jones (1967) discusses the several types of leishmaniasis caused by biologically distinct forms within each of the species *Leishmania dono-*

vani and *L. tropica.* He also suggests that speciation can occur in one generation, a view held by Bush (1974, 1975a, b) based on studies of herbivorous parasites.

The large size of many families of parasitic insects (Table 2.1) also supports the concept of high rates of evolution and speciation. Even though some parasitic taxa evolved much later than predatory taxa, families of parasites on plants are on average almost eight times larger than those of predators, and families of parasites on animals are over ten times larger. In both groups of parasites the tenth family is larger than the first ranked family of predators.

Concept 2. *Adaptive radiation is extensive and its degree of development in each taxon of parasites depends upon four major factors*:

(i) *the diversity of hosts* in the taxon or taxa being exploited (i.e. numbers of species and degrees of difference between species),

(ii) *the size of the host target* available to potential colonizers (body size, population size, geographical distribution),

(iii) *the evolutionary time available* for colonization of hosts,

(iv) *the selective pressure for coevolutionary modification* (i.e. for specialization).

White (1973:760) stated that an ecologist's way of interpreting evolution "is as a series of adaptive reactions on the part of living organisms to the ecological opportunities of the environment—a progressive filling of 'niches'." Thus, part of an understanding of the evolution of the large numbers of parasite species can be found in an understanding of why so many niches are available and what

26

influences the presence or absence of parasites in these niches. Some of the factors are treated in this section.

(2.i) *Diversity of hosts.* Adaptive radiation can be most extensive when many related species of host are available for colonization, particularly if the hosts within a taxon differ in an important way relative to the requirements of the parasites. Many examples could be given of a significant relationship between number of species in each host category and number of parasites that exploit members of that category. This has been shown for 12 families of fleas, six families of parasitic insects from four orders on mammals, birds, and insects (Price, 1979), and hispine chrysomelids on Zingiberales hosts (Strong, 1977a). It is also evident in the host records for mites on Lepidoptera provided by Treat (1975) and species of *Heliconius* butterfly on genera and subgenera in the host plant family Passifloraceae (Benson, Brown, and Gilbert, 1975). There are more species of leaf miner in the family Agromyzidae on large families of plants than small families (Price, 1977) (Table 2.3 summarizes data on these studies). These examples support

TABLE 2.3. Some relationships between number of hosts available and the number of parasites exploiting these hosts.

Parasites	Hosts	Variance accounted for (r^2)	Slope of linear regression (b)	Source
Agromyzid flies	Angiosperm families	0.61	0.05	Price (1977)
Heliconius butterflies	Passifloraceae genera & subgenera	0.92	0.38	Benson, Brown, & Gilbert (1975)
Hispine chrysomelids	Zingiberales families	0.86	1.47	Strong (1977a)
Mites	Lepidoptera families	0.96	0.24	Treat (1975)
Fleas	Mammals & birds	0.34	0.08	Price (1979)
Insects	Mammals, birds, & insects	0.99	0.07	Price (1979)

27

Eichler's (1948) rule, well known to parasitologists, that states (from Noble and Noble, 1976:504), "When a large taxonomic group (e.g. family) of hosts consists of wide varieties of species is compared with an equivalent taxonomic group consisting of few representatives, the larger group has the greater diversity of parasitic fauna."

At each successive trophic level the diversity of potential hosts increases and the opportunities for adaptive radiation expand, within the limits discussed below. Thus parasitic herbivores may be abundant, with many large families included (Table 2.1), but radiation has been even more extensive in the largest families of parasitic carnivores. The family Ichneumonidae is over three times larger than the largest family of primary parasites in the British fauna, and Townes (1969) estimates a total of about 60,000 species must exist in the world. Only at the fourth trophic level in a parasite food chain (e.g. obligate hyperparasites of parasitoids) may resources be so dispersed and hard to find that a trend towards generalization and reduced numbers of species might be expected (Darwin, 1872; Janzen, 1975).

The resulting diversity of parasitic organisms supported after the evolution of a new genus or species may be impressive. An extreme case is seen in the genus *Quercus* where even in the small biota of Britain the two species of oaks support at least 439 species of parasitic herbivores (Table 2.4). The number of parasitic carnivores supported by these herbivores must be even higher. The winter moth *Operophtera brumata* is known to support 63 species of parasite (Wylie, 1960) and microlepidoptera in the genus *Phyllonorycter* may support about 50 species of parasitic insect (Askew, 1975). It would certainly be an underestimate to state that the two British species of oak are the primary producers for a thousand species of parasite.

28

TABLE 2.4. Numbers of herbivores on oak in Britain and numbers of parasitic wasps (Hymenoptera) on one or two representative species of herbivore in the same taxon. If all parasites on herbivores were specific to one host, almost 10,000 species would be expected at the third trophic level based on oak. A more realistic estimate is that about 50 percent of the parasitic Hymenoptera are host specific (see Chapter 5).

Herbivore taxon	Number of Herbivorous Parasites	Source	Number of Parasites On Representative Herbivore Species	Source
INSECTS				
Thysanoptera	2	Morris (1974)	2	Lewis (1973)
Hemiptera	30	Southwood and Leston (1959)	1	Dupuis (1949); Southwood & Scudder (1956)
Homoptera	39	Morris (1974)	1	Personal observation
Microlepidoptera	81	Ford (1949)	49	Askew (1975)
Macrolepidoptera	112	Stokoe & Stovin (1944, 1948)	63	Wylie (1960)
Coleoptera	50	Walsh (1954)	5	Burns & Gibson (1968)
Hymenoptera				
Symphyta	13	Morris (1974)	6-12	Lejeune & Hildahl (1954); Underwood & Titus (1968)
Cynipidae	50	Darlington (1974); Askew (1975)	12-20	Askew (1961); (1975)
Diptera				
Cecidomyiidae	13	Barnes (1951, 1955)	5	Askew & Ruse (1974)
FUNGI	49	Murray (1974)	0	
TOTAL SPECIES	439		144-158	

(2.ii) *Size of host target.* The large number of parasites coexisting on oaks is possible partly because of the trees' large size. The many resources available for colonization and the species that exploit them have been described by Morris (1974).

The mean number of parasite species per host species in a family of plants is greater in trees than in shrubs and greater in shrubs than in herbs. This trend is illustrated in host records provided by Ford (1949) on the microlepi-doptera in Britain (Figure 2.1) and by Hering (1957) on leafmining Lepidoptera (Figure 2.1), Coleoptera, and Hy-menoptera in Europe. Similar relationships have been ob-served by Strong and Levin (1975, 1978) and Lawton and Schröder (1977). Among ectoparasites of birds (e.g. biting lice, Mallophaga, and feather ticks, Sarcoptiformes) equiva-lent relationships have been identified, with large birds having more regions of the body that are sufficiently distinct so that different parasites can colonize and maintain a com-petitive edge in each region (Dogiel, 1964). For example, in an extensive survey Foster (1969) discovered three mallo-phagan species that occupied two body regions in the orange-crowned warbler *Vermivora celata* whereas Dubinin (in Dogiel, 1964) found seven species on the larger bird *Ibis falcinellus* located in four distinct body regions. The survey by Theodor and Costa (1967) on ectoparasites in Israel also shows that small birds such as the Old World warblers (Sylviidae) and finches (Fringillidae) support a mean of one or less lice species per host species whereas larger birds like the plovers (Charadriidae) and falcons (Falconidae) support a mean number of about three lice species per host. A similar relationship can be seen in Bequaert's (1956) list of hosts of the louse flies (Hippo-boscidae) in the New World. Whitaker (1979) also found the trend among ectoparasites of insectivores and in some rodent groups, but not in bats, squirrels, and carnivores.

Host population size as well as the geographic distribu-tion of a species are equally important in determining the probability of a parasite reaching a new host and estab-lishing on it. Large populations and extensive range of a potential host must increase the chances of colonization by parasites as shown for insects on British trees (Southwood, 1961; Strong, 1974a, b) and pests of cacao (Strong, 1974c),

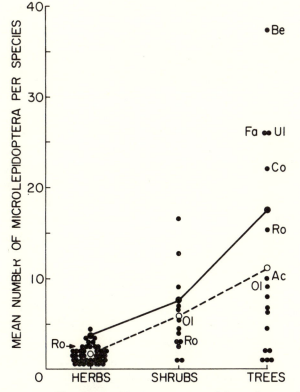

FIGURE 2.1. The relationship between size of host plant and number of parasitic herbivores per host species. Small closed circles (·) indicate the mean number of parasites per host species for microlepidoptera in Britain on each family of plants with a given form (host records from Ford, 1949). Open circles (o) and dashed line indicate general trend of increased number of parasites per plant species as plant size increases. Large closed circles (•) and solid line indicate the same trend for mean number of European leafmining Lepidoptera per plant genus in each family of plants with a given form (host records from Hering, 1957). Families represented by species falling into each of two or three plant form categories were subdivided as shown for the Rosaceae and Oleaceae. Ac indicates Aceraceae; Be, Betulaceae; Co, Corylaceae; Fa, Fagaceae; Ol, Oleaceae; Ro, Rosaceae; Ul, Ulmaceae. The large range in mean number of parasite species per host species of tree is discussed in Evolutionary Concept 2 (iii) (from Price, 1977).

31

fungi on trees (Strong and Levin, 1975), insects on sugarcane (Strong, McCoy, and Rey, 1977), insects in Britain on perennial herb species, annual herbs, woody shrubs, and monocotyledonous plants (Lawton and Schröder, 1977), and mites on cricetid rodents in general in North America, and mites on species in the genera *Peromyscus* and *Microtus* (Dritschilo et al., 1975).

Perhaps Kellogg (1913) was the first to regard hosts as islands available for colonization by parasites. Janzen (1968, 1973a) also points out that hosts can be regarded as islands, and they are therefore subject to the insights of the theory of island biogeography (MacArthur and Wilson, 1967) as Opler (1974) and Dritschilo et al. (1975) have demonstrated. Probability of colonization is influenced by island size that can be regarded as host size, host population size, or magnitude of host geographic range.

(2.iii) *Evolutionary time available.* Wilson (1969) envisioned four phases in the colonization of islands, the noninteractive, interactive, assortative, and evolutionary. Initially species are able to colonize uninfluenced by resident species, but eventually interactions are inevitable. The number of species on an island may increase by the assortment of species for improved coexistence, and ultimately more species may pack in through evolution of increased niche specialization. Such processes occur in parasite communities. Hopkins (1949) suggested that, after successful colonization, subsequent absence of ectoparasites on vertebrates may be caused by competition between different groups of parasites. He noted a form of competitive complementarity between Mallophaga and Anoplura among the rodents: "The families and genera which are heavily infested with Mallophaga seldom or never have Anoplura" (1949:430). Similar relationships probably exist between

32

fleas, lice, and mites (Thompson, 1938; Brinck, 1979), and
Halvörsen (1976) reviews many cases in the literature on
parasitic worms (but see also Chapter 6).

In Evolutionary Concept (2.ii) trees were shown to support on the average more parasitic herbivores than herbs
(Figure 2.1). However, the range in mean number of parasites per species of tree is extreme and is not accounted for
by the size of the host individuals. Some of this range can
be explained by the relative evolutionary opportunity provided by each species of tree. Common trees that have
existed in a region through extensive spans of time have
high numbers of parasites whereas recently available hosts
with a restricted range have small numbers (Southwood,
1961). Although this conclusion has been criticized (Strong,
1974a, b; Claridge and Wilson, 1978), it is clear that specialists are accumulated by a host in evolutionary time
(Southwood, 1977; see also Lawton and Price, 1979).

Once parasites have colonized a host, divergence of host
stock and eventual speciation leads to divergence of the
parasites, and, depending upon the time involved, results
in formation of a new host race or new parasite species.
Thus parasites can be extremely useful in unravelling the
phylogenetic relationships of their hosts (e.g. Jordan, 1942
on Siphonaptera; Harrison, 1914; Metcalf, 1929; Hopkins,
1942, 1949; Clay, 1950, on Phthiraptera; Manter, 1966 on
Trematoda; Lindquist, 1969 on Acarina; Saville, 1975 on
Uredinales), including the identification of sibling species
in the host group when the parasites diverge more notice-
ably than their hosts (Mayr, Linsley, and Usinger, 1953;
Mayr, 1963). Such diverse groups as beetles on pines, termit-
ophiles in termite colonies, and ciliates on turbelarians
have provided clues to inconspicuous differences between
hosts. This process has been observed sufficiently often to
be formalized into a rule by Fahrenholz (cited by Noble

and Noble, 1976): Common ancestors of present-day parasites were themselves parasites of the common ancestors of present-day hosts. Degrees of relationship between modern parasites thus provide clues to the progenitors of modern hosts. The more restricted Fuhrman's rule (Dogiel, 1964), that each order of birds has its particular cestode fauna, implies a similar relationship between the evolution of host and parasite.

(2.iv) *Selective pressure for coevolutionary modification.* The importance of interaction between host and parasite in the evolution of both has been stressed by numerous authors (e.g. Brues, 1924; Flor, 1956, 1971; Dogiel, 1964; Ehrlich and Raven, 1964; Day, 1974; and authors in the symposium edited by Wallace and Mansell, 1976). The stepwise coevolutionary process results in extreme specialization and complex defense mechanisms (e.g. Whittaker and Feeny, 1971; Rhoades and Cates, 1976; Wakelin, 1976; Deverall, 1977). As suggested in Ecological Concept 2 and Evolutionary Concept 3, specialization is likely to increase the rate of speciation that may occur in both host and parasite. Indeed, as Atsatt (1973) pointed out, parasites may have increased the adaptive potential of their angiosperm hosts enabling the evolution of heterotrophic species, including the parasitic flowering plants. The role of symbiotic microorganisms in the evolution and specificity of parasites will be discussed further in Chapters 3, 5, and 7.

The importance of coevolutionary pressure may be illustrated in the relationship between agromyzid leaf miners and their plant hosts. Families containing species that are biochemically distinct have relatively high numbers of specialist agromyzids, whereas families of low chemical diversity have fewer, more generalized species. This is seen

34

particularly clearly when host specificity is compared for agromyzids on Umbelliferae and Graminae. The Umbelliferae is composed of aromatic plants that produce a diverse array of essential oils and related resins (Hegnauer, 1971) with a large number of pharmaceutically interesting species (Heywood, 1971). The Graminae, in contrast, show low chemical diversity compared to the Umbelliferae, and a higher species per genus ratio (Table 2.5) indicating greater taxonomic uniformity within the family.

TABLE 2.5. The size of the families Graminae and Umbelliferae (from Clapham, Tutin, and Warburg, 1957), the species per genus in each family, the number of genera attacked by agromyzids in Europe, and the number of species of agromyzid found on members of these families (from analysis of keys in Hering, 1957). Note that the Graminae is a larger family than the Umbelliferae, but there are fewer species per genus in the latter group indicating relatively greater taxonomic diversity.

	Graminae	Umbelliferae
Genera in World flora	c. 600	c. 200
Species in World flora	c. 8-10,000	c. 2,700
Genera in British flora	59	43
Species in British flora	170	69
Species per genus in British flora	2.88	1.60
Genera attacked by agromyzids in Europe	49	42
Species of agromyzids attacking each family in Europe	35	61
Species of agromyzids on each genus attacked (mean)	4.32	2.49

Although the Umbelliferae is a smaller family than the Graminae, many more agromyzid species attack members of the family in Europe (61 species on Umbelliferae, 35 species on Graminae). This is apparently because the chemical diversity of potential hosts within the Umbelliferae has forced specialization of the parasites. Eighty-two percent of the species of agromyzid attack only one genus each (Figure 2.2). (Unfortunately, in Hering's (1957) keys used to

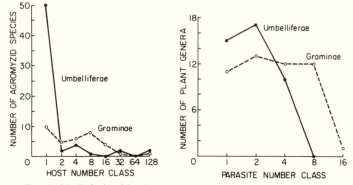

FIGURE 2.2. Relationships between agromyzid flies and their hosts in the families Umbelliferae and Graminae in Europe (host records from Hering, 1957). (Left) Number of agromyzid species in each class indicating the number of hosts utilized. (Right) Number of plant genera in each class indicating the number of parasites per genus. Classes are arranged on a logarithmic scale with the first number in the class indicated (from Price, 1977).

obtain these data, plant species were grouped under genera so parasite specificity per host species cannot be obtained.) By contrast only 29 percent of agromyzids attack one genus in the grasses. In addition, there are fewer agromyzids on each genus of Umbelliferae than on Graminae (Figure 2.2). No more than 7 species occur on any one genus in the former family and 15 genera have only one parasite species. When there are few parasite species per host, coevolution can proceed rapidly since adaptive reactions need not be compromised by conflicting adaptations in response to other parasites exerting different selective pressures. Reciprocal responses by the parasite to the host can also be better tuned when only one host is utilized, as demonstrated in the artificial evolutionary arena conceived by Pimentel and Bellotti (1976). When "parasite" populations were exposed to single host resistance factors, the populations rapidly evolved the capacity to tolerate them (in 7 to 10 generations), but when all six resistance factors were combined, the parasite population showed no evolved tolerance in 30 generations. Thus in systems of a single host and a single parasite adaptations for specialization are reinforced, isolation of populations becomes more likely, and speciation is more rapid. As Mayr (1963:462) expresses it, "Host specificity is thus an ideal prerequisite for rapid speciation," a view also supported by Ross (1962). The end result is a comparatively large number of specialists attacking the Umbelliferae whereas many more generalists attack the Graminae.

The degree to which specialization is demanded is a potent force in adaptive radiation. Szidat's rule (Eichler, 1948) states that the more specialized the host group, the more specialized are its parasites; and conversely, the more primitive or more generalized the host, the less specialized are its parasites. Hence the degree of specialization may serve as a clue to the relative phylogenetic ages of the hosts (as stated in Noble and Noble, 1976:504). Predators must remain generalized and radiation in any taxon has been unimpressive (Table 2.1).

Plant parasites show varying degrees of radiation depending on the intimacy of their association with the host. Plant bugs (Miridae) are mobile, relatively large ectoparasites, although immature stages spend much time on a single host. In the British fauna 186 species have been identified (Table 2.1). The much smaller plant lice (Aphididae) are more sessile, more intimately associated with the host and have undergone more extensive radiation: there are 365 species in the British fauna. Some are gall makers, and production of a distinctive gall is always associated with a very specific relationship between host and parasite (Eastop, 1973). The largest families of plant parasites are predominantly endoparasitic. The weevils (Curculionidae) as larvae mine in leaves, under bark, in shoots, roots, fruits, seeds, or feed in rolled leaves. They number 509 species. The most highly coevolved parasitism occurs in the gall-forming endoparasitic flies Cecidomyiidae, which are also the most numerous herbivorous parasites with 629 species in the British fauna (Table 2.1). It is not clear why members of the Cynipidae, which also form galls, are not as well represented in this fauna.

Specialization in parasites on animals should be more highly developed than in parasites on plants. Although animals as a group are chemically more similar than plants, herbivores have a greater diversity of places to live than plants and show strong and diverse behavioral, phagocytic, and immune defenses against parasites. Thus, finding and living with hosts seems to demand a greater number of adaptations per species for parasites of animals than for parasites of plants, and therefore a narrower host range. Members of the largest families in the British insect fauna are parasites of herbivorous parasitic insects: the Ichneumonidae with 1938 species, the Braconidae with 891 species, and the Pteromalidae with 649 species.

Probably the degree to which species of host and parasite are coevolved and the proportion of the genome devoted to coadaptation could not have been appreciated without extensive breeding of plants for parasite resistance. Flor (1971) and Day (1974) provided strong evidence from plant breeding experiments for a gene-for-gene relationship between the plant host's resistance and the associated parasite's virulence. After prolonged coevolution where the stepwise process has escalated defenses many times, many such complementary gene-for-gene pairs must exist between parasite and host. This complex of genes coadapted to counterparts in the host, like the closed variability system described by Carson (1975), must be maintained by inversions, development of supergenes, or by cloning, the last so often seen in parasitic organisms (see Ecological Concept 1 and Chapter 4). Specificity of parasites is discussed in more detail in Chapter 5.

Previously I have listed mobility of hosts as an important factor in adaptive radiation since high mobility reduces isolation between populations and low mobility reinforces it (Price, 1977). The high mobility of birds and some mammals would then account for the apparently slow evolution of parasites utilizing these hosts (e.g. lice, Phthiraptera, and fleas, Siphonaptera) noted by several authors. However, since then I have not found evidence that the lice or fleas have speciated less than other groups of parasites (Price, 1979, see also Chapter 5), making it less certain that host mobility is a major factor in adaptive radiation. Marshall (1976) emphasized the importance of mobility of the parasites themselves, although he recognized as an additional factor the significance of physical isolation between parasite populations that would be decreased by high host mobility. Also, other factors influencing parasite diversity, considered by Whitaker (1979) such as use of burrow systems, nesting in holes, and social roosting may mask any effect that host mobility has on parasite species richness.

Concept 3. Types of speciation other than through geographic isolation are at least as important as allopatric speciation. As Mayr (1963) predicted, arguments in favor of sympatric speciation have continued to appear since he soundly criticized the concept. This is not so much because of continued ignorance of the past literature as Mayr claimed but because botanists and entomologists in particular have not been convinced that only one method of speciation can account for all cases (White, 1968, 1974). A growing number of evolutionary biologists would subscribe to the view held by Kinsey (1937:5): "Discussions of evolutionary problems usually fail to take into account the varying constitution and behavior of species among different groups of plants and animals. . . . The qualities of interbreeding populations among such diverse things as, for instance, strong-flying birds, parasitic insects, world-circling primates, microscopic fungi, bacteria, slow-fruiting angiosperms and crustacea of the ocean plankton are so various that no single, simple explanation, or any theoretic ideal, can account for the evolution of them all." A parallel view has been expressed by White (1973:768): "The extent of vagility and the population structure of the group seem to play a most important role in determining what kinds of cytogenetic changes can establish themselves in populations and hence play a role in speciation." There are probably several, or even many routes by which species can be formed (see White, 1968; Scudder, 1974; Bush, 1975b), and sympatric speciation is one important means (Stebbins, 1964; Grant, 1966; Maynard Smith, 1966; Spieth, 1968; Thoday, 1972; Bush, 1975b; Endler, 1977). Thoday (1972) remarks that heterogeneity of habitats is often underestimated and that a population is exposed to disruptive selection if more than one phenotype has optimal fitness and intermediate phenotypes have lower fitness. He concludes that some species differences are maintained by disruptive selection, that also maintains differences between local populations, and that such selection can result in isolation between populations. Pure allopatric isolation is not the only means through which speciation occurs (see also Bush, 1975b).

Ross (1962) emphasized the importance of host shifts that isolate sympatric populations of a parasitic species, and Mayr (1963:460) stated that host races of phytophagous animals "constitute the only known case indicating the possible occurrence of incipient sympatric speciation." In subsequent publications (Mayr, 1970, 1976) Bush's arguments were becoming more convincing to Mayr. If host race formation can lead to speciation of phytophagous parasites, it can also be important among animal parasites (see White, 1978 for a critical evaluation of the evidence). Rapid evolutionary rates and such extensive adaptive radiation as seen among parasites are not easily explained by allopatric speciation over extended time periods. The large numbers of sibling species are equally hard to explain by this model. Bush (1975b) concluded that host races of plant parasites evolved sympatrically and that these were undoubtedly the progenitors of the many sibling species so often found to be sympatric on different hosts.

Bush (1974, 1975a) suggested that the establishment of new host races may require only minor alterations in the genome. His basic model accounts for speciation involving only two alleles at each of two loci: one locus controlling host selection and the other controlling survival in the host. One allele at each locus carries these traits adapted to host species *A*, and the other alleles enable the parasite to discover and exploit host species *B* (see Chapter 4 for details). Disruptive selection works against individuals with alleles for selection of one host and survival in the other.

If host species *A* and *B* have different phenologies and the parasites adapt to these differences, reproductive isolation is reinforced. Bush (1975a) provided several examples of this allochronic isolation involving host shifts that can occur through "a narrow window in space and time." For example, on Mount Shasta in California a host shift of the fruit fly *Rhagoletis indifferens* from bitter cherry *Prunus emarginata* to introduced domestic cherry *P. avium* can occur only at about 5,000 feet during the last two weeks of July, whereas the fly occurred from sea level to 9,000 feet during the period from May to October. Tauber and Tau-

ber (1977) provide evidence of sympatric allochronic speciation in a small predator and propose a simple genetic model
depicting the mechanisms involved (see also Tauber, Tauber, and Nechols, 1977). Triggering of reproductive activity
in the parasite may also be initiated by the host as in the
rabbit flea *Spilopsyllus cuniculi* and hare flea *Cediopsylla
simplex* (Rothschild, 1965; Rothschild and Ford, 1964,
1973), insects parasitic on others (Salt, 1941; Lees, 1955;
Wigglesworth, 1965), protozoa in insects (Malavasi et al.,
1976; Cleveland, 1960) and other hosts (El Mofty and
Smyth, 1960), and possibly among blood-feeding lice on
birds (Foster, 1969).

From the gene-for-gene hypothesis Day (1974) predicted
the number of host races that a set of resistant varieties
will select for. With 19 genes for resistance in apple, for
example, each of which may have two phenotypes, resistant
or susceptible, there would be 2^{19} or 524,288 races of a
parasite adapted to exploit fully the range of apple varieties.
This may seem an extreme number, but we have every reason to infer that a parasite must be closely attuned morphologically, physiologically, and biochemically to the host,
and such extensive race formation may be necessary and
realistic. Once the races have differentiated in this way,
subtle ecological or temporal isolation could easily promote the independent differentiation of populations.

Since most parasites are small, usually with narrow
tolerances to environmental factors, they are susceptible to
minor spatial or temporal change. When tolerances are
narrow, slight differences between habitats may cause isolation where habitats may be only 100 meters apart. If repro-
ductives or dispersing individuals live only a few days, a
week's difference in phenology may prevent gene flow between populations. Many more ecotones exist for parasitic species, and it is in these intermediate and changeable

zones that Stebbins (1974) sees the cradle for rapid evolution. For such small, short-lived, precisely adapted organisms as parasites, evolution will operate in miniature—in short times, in small spaces, but with impressive results.

Based on the six concepts outlined in this chapter, some trends can be expected in a parasite food chain based on plants. Ecological conditions at each successive trophic level will be characterized by (1) more patchily distributed hosts and (2) less predictable resources. Characteristics of the species will tend toward (3) smaller size, (4) shorter life cycles, (5) more specialization (i.e. lower ranges of tolerance), (6) greater population fluctuations, and (7) greater isolation between populations. Evolutionary consequences will be (8) higher evolutionary rates and (9) more extensive adaptive radiation.

CHAPTER THREE

Non-Equilibrium Populations
and Communities

NON-EQUILIBRIUM POPULATIONS

Equilibrium in a population exists when the birth and death rates are equal; net growth rate is zero (MacArthur and Wilson, 1967; May, 1973). Such an ideal prevails only momentarily in stochastic environments, but here equilibrium will exist when population size fluctuates with a steady average variance around an average population size (May, 1973) (Figure 3.1a). Stable populations will have low variance, and unstable populations will have high variance. The concepts of equilibrium and stability dominate theoretical ecology today as they have dominated this science and general biology since their beginnings.

A population in a non-equilibrium state may be defined as one that does not fluctuate within a typical probability range around an average population size. Steady population growth following colonization with rapid extinction (Figure 3.1b) yields no indication of an equilibrium. Population size during tenure of a site in which carrying capacity changes stochastically also reveals no equilibrium (Figure 3.1c). In the absence of an identifiable equilibrium the terms stable and unstable as used above become derelict. Where "stable" non-equilibrium behavior of populations has been identified, the term has been more closely allied to the concepts of persistence and resilience in populations (e.g. Beddington, Free, and Lawton, 1976) due to the existence of

44

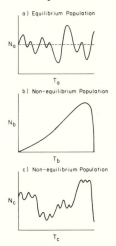

FIGURE 3.1. Examples of equilibrium and non-equilibrium populations. (a) Population size *(Na)* varies through time *(Ta)* about an equilibrium number (dashed line). (b) Population becomes established, increases rapidly and crashes to extinction. (c) Population fluctuates about an equilibrium determined by a carrying capacity that frequently changes.

limit cycles or chaotic behavior within well-defined boundaries.

Ecological Concept 3 in Chapter 2 stressed the need for a detailed within-patch approach to understanding parasites rather than the classical treatment of total population over many patches. Indeed it is apparent that the dynamics of the total population of a parasitoid and its host cannot be understood without consideration of the within-patch dynamics (Beddington, Free, and Lawton, 1978). During the lifetime of an individual, parasitic organisms will normally experience conditions in one or at the most two patches, and their adaptive strategies must be attuned to life within a patch. By contrast, a predator or a browser, which may have food resources just as patchily distributed,

45

searches many patches and optimizes its allocation of time and energy by testing for, and exploiting, the most profitable. These animals share with the theoretical ecologist an interest in the overall pattern of patches and the dispersion of food within these patches, and the theory of optimal searching strategies is well developed (e.g. Pyke, Pulliam, and Charnov, 1977; Emlen, 1973; Pianka, 1974; Royama, 1970). Achievement of equilibrial states is the preoccupation of the modeller. But the theory of non-equilibria is poorly developed. It has frequently concerned coexistence of competitors and not species on different trophic levels (e.g. Skellam, 1951; Hutchinson, 1953, 1965; Levins and Culver, 1971; Horn and MacArthur, 1972; Slatkin, 1974). The subject has not received the attention it deserves, perhaps mostly because of the mathematical hurdles in treating non-equilibrium states. However, there is a recent increasing attention to modelling non-equilibrium situations that relate to the dynamics on two trophic levels.

Caswell (1978) draws the line nicely between equilibrium and non-equilibrium in the history of conceptual ecology and points out that acceptance of non-equilibrium in populations is implicit in the concepts of ecological succession and the dynamic equilibrium of communities. Several recent models depict the within-patch disequilibrium in a population, although global equilibrium reigns both in single species systems (Gurney and Nisbet, 1978) and in predator-prey systems (Hastings, 1977, Zeigler, 1977). Such global equilibrium is also observed in parasite-host systems where the proportion of patches occupied, that is the proportion of host individuals infected, remains remarkably stable in some instances. Anderson (1979) cites the particularly impressive examples of *Schistosoma haematobium* (Trematoda) in man in Iran (Rosenfield, Smith, and Wolman, 1977) and *Pomphorhynchus laevis* (Acanthocephala) in both the definitive and intermediate hosts (Kennedy and Rumpus, 1977). But the mechanisms resulting in such steady equilibrium probability distributions for host populations may involve local disequilibrium in parasite populations. Indeed, the highly aggregated distribution of parasites in hosts is a crucial aspect of the general models of interactions between parasite and host by Crofton (1971), May (1977a), Anderson and May (1978), and May and Anderson (1978). Such overdispersed attack patterns are also modelled for schistosomes (May, 1977b) and for parasitoid-host systems (May, 1978a). Anderson and May (1978) reviewed the empirical data on parasite dispersion in hosts and concluded that many parasites are highly overdispersed. That is, hosts accumulate parasites at very different rates; many hosts support populations well below their carrying capacity, and perhaps extinction rates are highly variable within the host population.

Emphasis on within-patch dynamics is not a radical departure from major currents in ecology but simply a shift to the detailed study of the individual unit of resources instead of the mean condition of resources over many units. Huffaker's (1958) classic study on orange mites graphically portrayed the within-patch dynamics that were certainly non-equilibrial, but the preoccupation was with total arena phenomena, not with population dynamics on a single orange.

Emphasis on non-equilibrium within patches also does not necessarily conflict with parasite-host models that identify mechanisms resulting in stability of host populations. Host populations are the global aspect of many local and intimate interactions. Also the models tend to concentrate on pure aspects of interaction whereas this chapter will include consideration of disruptive influences on the interaction. May and Anderson (1978) dwell on destabilizing processes in interactions between parasites and hosts, and Anderson (1979) states that only small changes in the probability of parasite transmission, which could result from climatic change, may result in epidemics of a disease. And, as Levin (1976) says, biological interactions exaggerate heterogeneity, so with each interactive cycle the spatial arrangement of patches should be modified, a characteristic yet to be modelled. The frequency of disruptive influences on a stable parasite-host interrelationship has not been measured empirically. I claim in this chapter that disruption in parasite-host systems is common. Nobody disclaims the existence of disruptive agencies, but we may well disagree on their frequency until many field studies yield the relevant information. Levin (1976) provides many examples from the literature largely on nonparasites.

The thesis to be defended in this section is that within patches parasite populations generally exist in non-equilibrium conditions. The central generating force for non-equilibrium appears to be the complexity of biotic associations characteristic of parasitic ways of life (Figure 3.2). Population dynamics of the parasite may depend upon the overlap of several other organisms and their dynamics in time and space making resources patchy and patches ephemeral as in the chigger-borne rickettsiosis example given in Chapter 2. Small, short-lived patches mean for the parasite that there are low probabilities of colonization and high probabilities of extinction and no mechanism for equating the two, with most individuals living in non-equilibrium populations (Figure 3.2).

Complexity of biotic associations among parasites takes several forms:

(1) There is the obligatory relationship with another living organism, the host that has its own dynamical properties

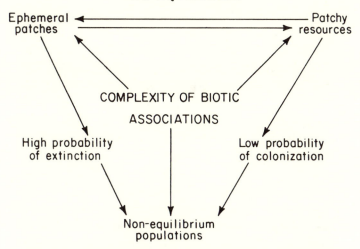

FIGURE 3.2. Patch dynamics in parasite populations: the general case. The term patch dynamics was coined, and its importance stressed, by Thompson (1978) and Pickett and Thompson (1978).

in space and time, its own relations, direct and indirect, with other species. The host may be a parasite itself in which case at least three organisms in a trophic hierarchy become tightly linked.

(2) Many parasites are heteroecious, involving two or three hosts in any one life cycle; each host has unique requirements and properties.

(3) Obligatory mutualism between parasites and microorganisms is commonly observed, the latter playing an important role in mediating the interaction between parasite and host. Microorganisms are associated with arthropods feeding exclusively on blood: ticks, bugs, lice, and louse flies (Buchner, 1965). Fungi are mutually associated with many plant-feeding insects: wood wasps, gall midges, bark and ambrosia beetles (Batra and Batra, 1967; Graham, 1967), and virus-like bodies with

49

parasitic wasps (Vinson and Scott, 1975; Stoltz, Vinson, MacKinnon, 1976). Thus conditions must be suitable for the mutualist as well as the parasite.

(4) The presence of one parasite frequently predisposes the host to access by another parasite. Root-knot nematodes expose the host to *Fusarium* wilt making the plant susceptible to a foliage pathogen, *Alternaria tenuis* (Powell, 1971). As Powell points out, we cannot estimate how many multiple interrelationships like this exist in nature or their impact on the host species, but judging from the literature on plant pathogens alone, we must expect such complex associations to be very common and significant in the ecology of many hosts. As Cohen (1976:256) says, "most parasitized people are miniature ecosystems of multiple parasites which interact."

(5) All these relationships, since they are between living organisms, are subject to coevolutionary change. This may proceed rapidly and independently in different patches as argued in Evolutionary Concept 1.

The term *biotic association* has been used in Figure 3.2 in the classic phytosociological sense that the ecologies of species are inextricably linked. The only limitation on this use is that it applies from the parasite's point of view and not from the host's. The ecological niche of a parasitic nematode or mite on a scolytid beetle, which is mutually associated with a fungus, and which together exploit a tree host, will always include the beetle, fungus, and tree, but the ecological niches of the beetle and the tree do not include the nematode or mite.

Such complexity of biotic associations inevitably leads to extreme patchiness of resources because the parasite can only survive where its niche overlaps that of its host. Only a segment of the host population is probably available to a

50

parasite species, and this segment is probably patchily distributed. Recent literature has stressed the patchy nature of plant distribution (Pickett, 1976; Feeny, 1976; Rhoades and Cates, 1976; Collins, Mitchell, and Wiegert, 1976; Grubb, 1977). Even in climax communities gaps become available frequently for colonization and create patchiness (Grubb, 1977). Coles and Fowler (1976) also allude to a family structure of trees within a stand; adjacent trees were closely related and exhibited inbreeding depression, while trees only 100 meters apart were apparently unrelated. Polymorphism is common in forest trees (see Stern and Roche, 1974 for review). Thus for highly coevolved parasites there exists an appreciable genotypic patchiness among hosts that tends to dilute the apparently available resources.

In response to this genotypic patchiness in climax communities parasites show the expected specialization, even down to differentiated demes on each host individual. Edmunds (1973) found that specialized demes of the black pineleaf scale *Nuculaspis californica* were adapted to individual host trees, with progeny surviving best on the parent tree. The pressures for specialization in a deme to a single host individual were thought to be so great that extinction of the deme would occur when the host tree died (Edmunds and Alstad, 1978). Riom and Fabre (1977) also noted the strong tendency for inbreeding of scale insects on a single tree, with similar consequences no doubt. Where such specialization is demanded it is not surprising that some parasitic herbivores, like the fall cankerworm *Alsophila pometaria* have become flightless and parthenogenetic (Mitter and Futuyma, 1977). Several clones of different genotypes are associated with oaks, two with maple, and two with mixed forest, and dramatic shifts in clone frequency within 100 meters occur as host species change (Mitter et al., 1979; Schneider, 1980). The clones differ, for example, in phenology, which correlates well with the phenologies of their major hosts.

Many other qualitative differences between hosts render a significant fraction of the population unavailable to a species of parasite. Thin-stemmed members of a plant population were not attacked by stem-boring insects (Rathcke, 1976; Mook, 1971) or gall formers (Mook, 1967, 1971); many small plants in dense stands were not attacked by a parasitic herbivore that utilized large isolated plants (Thompson and Price, 1977). The age specificity of pathogens on humans and other vertebrates provides many instances of the influence of host quality on parasite invasibility. Indeed, suitable hosts may be rare for a particular parasite species or genotype in an apparently plentiful supply of host individuals, and our ability to identify the size and suitability of resource patches is probably completely inadequate at present (see also Collins, Mitchell, and Wiegert, 1976).

Complex biotic associations have many weak points. Conditions for any one member of the symbiotic complex may become intolerable and the association collapses. Thus, with every increase in complexity not only will patchiness increase, but viability of patches will decline. Complexity results in non-equilibrium conditions in time and space in parasite systems. May (1973) also concluded that complexity decreases stability in model systems, and Pimm and Lawton (1977) and Beddington and Hammond (1977) found that as food chains lengthen they become much less stable.

Ephemeral patches may be created by a variety of processes. Host populations will change dramatically within a patch, influencing the ability of a parasite to exploit that population, probably in a density-dependent manner. Conditions for the host are likely to improve and then decline through ecological succession, imposing a longer term pattern of host availability. Late colonizers of a patch may overcome original residents. The host may be vulnerable to parasite attack only briefly in its life cycle. This renders patches particularly ephemeral in hosts with highly synchronized life cycles such as many plants and animals in temperate regions, deserts, and the dry tropics.

Some examples will support the general statements made here, and further examples may be found in the section on non-equilibrium communities.

The complexity of biotic associations, the patchiness, and rapid change in patches typical of the epidemiology of chigger-borne rickettsiosis were covered briefly in Chapter 2. These relationships are summarized in Figure 3.3, which takes the same form as the general case in Figure 3.2. An important additional ingredient in the epidemiology, not mentioned in Ecological Concept 3, is the population dynamics of the small mammals that act as hosts to chiggers and *Rickettsia*. The relationship between chiggers and rats is particularly critical to the maintenance of *Rickettsia* populations since the pathogen passes from one generation of chiggers to another by transovarial transmission (Traub and Wisseman, 1974). Small mammal populations will respond to increases in food availability, such as large seed production in some years, and population pressure will cause movement of individuals out of enclaves over more inhospitable terrain. Synchronized mass movements of small mammals are well known and are discussed in detail by Hershkovitz (1962) for South American cricetines. These so-called ratadas or rat-plagues no doubt contribute substantially to the dynamic properties of chigger-borne rickettsiosis. Expansion of a population with chiggers and *Rickettsia* may carry the parasites to new locations that had harbored suitable rodents free of these parasites. Other in-

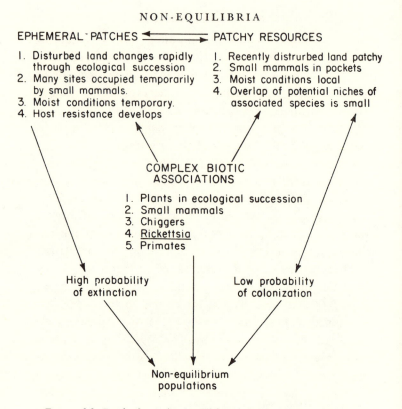

EPHEMERAL PATCHES ←————→ PATCHY RESOURCES

1. Disturbed land changes rapidly through ecological succession
2. Many sites occupied temporarily by small mammals.
3. Moist conditions temporary.
4. Host resistance develops

1. Recently distrurbed land patchy
2. Small mammals in pockets
3. Moist conditions local
4. Overlap of potential niches of associated species is small

COMPLEX BIOTIC ASSOCIATIONS

1. Plants in ecological succession
2. Small mammals
3. Chiggers
4. Rickettsia
5. Primates

High probability of extinction

Low probability of colonization

Non-equilibrium populations

FIGURE 3.3. Patch dynamics in *Rickettsia* and chigger populations.

dividuals may colonize new sites unsuitable for chiggers. Contraction of populations may occur rapidly and the pattern of infective patches may have changed dramatically within a few weeks. Thus the population dynamics of the food plants influence the population dynamics of the small mammals that in turn affect chigger populations and *Rickettsia* populations. Since even the plants may exist in a patch for only a few years because of succession and within this time seed production probably varies considerably, it is hard to conceive of conditions that would lead to equilibrium populations at any of the trophic levels, but least of all among the parasites.

Similar scenarios are observed in the ecology of viruses such as those causing yellow fever and tick-borne flavivirus encephalitis where birds and mammals act as reservoir hosts; man is also involved, and transmission is by arthropod vectors (cf. Fenner and White, 1976).

The more extensive the studies on a parasite species have been, the more evident are forces resulting in non-equilibrium populations. In *Schistosoma* species that utilize mammals as one host and snails as the other, our perception of the environmental grain in which they exist becomes increasingly coarser as knowledge accrues. The snail hosts are largely aquatic with some living in large bodies of water and others in small pools and temporary ponds (Berrie, 1973). Patchiness in the watery environment is extreme. Rain repeatedly creates new sites available for colonization. Intermittent flooding generates new patches, the destruction of others, and permits brief but considerable movement of snail hosts and free-living stages of the parasite: the miracidia and cercariae. Dry spells cause concentration of snail hosts and mammal hosts, destruction of many pools, and the creation of pools in river beds. Added to this patchiness is the poor dispersal ability of snail hosts even along major bodies of water such as the Nile (Berrie, 1973). As Berrie points out, snails are hermaphroditic and self-compatible, so a new population can be founded by a single individual. Such sedentary animals are likely to speciate parapatrically (Bush, 1975b), and clinal variation within a species and between species is apparent making the systematics complex and the resources available to the parasitic schistosomes patchy and highly variable between patches. Of snail populations in the genus *Bulinus* Berrie (1973:176) says, "Ecological and genetic factors are constantly interacting so that individual populations may develop quite distinctive features."

Snail populations therefore present enormous variety to parasitic schistosomes, and, as we should expect, strains of the parasite have evolved to exploit different hosts—"each snail is associated with its own strain of parasite" (Berrie, 1973:177)—and different locations of the same host (see Wright and Southgate, 1976). Where a schistosome species attacks more than one host, a population has been found to be polymorphic for infectivity and the hosts polymorphic for susceptibility. The snail *Biomphalaria glabrata* was shown to have seven susceptibility types using juvenile and adult snails and two strains of *Schistosoma mansoni* (Richards, 1976). Natural populations of this snail have been found with infection frequencies ranging from zero to 100 percent (Richards, 1976).

Added to the complexity of the snail host population structure and the variation in host susceptibility and parasite infectivity is the probability of hybridization between *Schistosoma* species (see Wright and Southgate, 1976). Such hybrids may inherit infectivity for both parental hosts, have higher infectivity in these hosts, or even prove infective to a completely new host species. In each case the hybrid schistosome may greatly expand its range and host shifts may be achieved by hybridization. Hybrids of *Schistosoma mansoni* and *S. rodhaini* were able to utilize *Biomphalaria alexandrina* as a host utilized by neither parent. In one group of *Schistosoma* all interspecific crosses have yielded viable hybrids, and, even though apparently less fit than parents in the parental milieu in some cases, their brief occurrence in nature with parental populations may be sufficient to yield significant new evolutionary pathways in subsequent generations, and in new environmental settings. Frandsen (1975) found greater viability of hybrids between sympatric strains of *Schistosoma intercalatum* and *S. haematobium* than viability of progeny from crosses of allopatric populations of *S. intercalatum*. The genetic state of schistosome populations, even though only crudely understood, is obviously very dynamic and in evolutionary terms potentially explosive. Such an open genetic system is reminiscent of those found in parasitic flowering plants like *Euphrasia* (Yeo, 1978) and *Orthocarpus* (Atsatt, 1970).

A close look at the mammalian hosts would yield another set of factors generating patchiness and non-equilibrium conditions. Susceptibilities and immune responses among primates are variable (Kagan and Maddison, 1975) and the conflict between parasite and host is sufficiently intricate (see Smithers and Worms, 1976) to select surely for host race formation. Ecologies of hosts are different enough to cause ecological isolation between populations of schistosomes (Wright and Southgate, 1976). Host populations, as discussed in the *Rickettsia* example, wax and wane in space and time. The overlap of all essential resources for schistosome parasites, with each resource in a dynamic state, will be patchy and brief, evanescent, and non-equilibrial (see Figure 3.4 for summary).

Many early successional plants, which are highly patchy in distribution, provide resources for parasites over a very short period of time. For example, the wild parsnip *Pastinaca sativa* overwinters as a rosette and then in early summer in rapid succession it elongates the stem, produces cauline leaves, inflorescences, green fruits, and finally dry fruits. Each herbivorous parasite is specialized in its attack on a particular stage: the leaf miner *Eulia fratria* on young leaves, larvae of the moth *Depressaria pastinacella* on young inflorescences. Young leaves were attacked by the leafminer only from June 1 to June 29 after which time no young leaves were available (J. N. Thompson, personal commu-

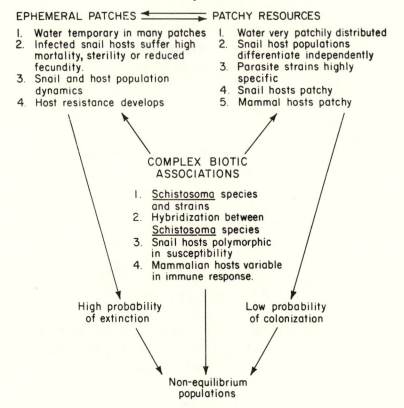

FIGURE 3.4. Patch dynamics in *Schistosoma* populations.

nication). Inflorescences remained suitable for feeding by *D. pastinacella* larvae only until July 13 (Thompson and Price, 1977), but oviposition of eggs was possible for a much briefer period while inflorescences were unopened. The plant occurs in early succession starting from disturbed ground, and within such disturbed areas it is very patchy with patches changing dramatically in location and size. Consequently, herbivores experience enormous shifts in carrying capacity from year to year (Thompson, 1978). Even within a year, resources for any one parasite are ephemeral. The dynamic pattern and quality of resources between seasons and within seasons leads to a parasite species normally existing in a non-equilibrium state within local patches.

The ephemeral, patchy nature of resources may also be caused by the activity of hosts. This is seen on a miniature scale in the interaction between the algal-bacterial mat in the effluent channels of the hot springs, the brine fly *Paracoenia turbida* (Diptera: Ephydridae), and its parasitic mite *Partnuniella thermalis* (Acarina: Protsiidae), described by Collins (1975) and Collins, Mitchell, and Wiegert (1976). As the algal-bacterial mat grows, it tends to block the flow of hot water, and small dams become cool enough for oviposition by the brine fly. Brine fly larvae convert the dam into a "stagnant soup" unattractive to adults, weaken the dam, and eventually cause its disintegration. Patchy, ephemeral host resources confront *P. thermalis*. Adult mites lay eggs in the algal-bacterial mat and larvae wait on algal islands for brine fly adults. Mite larvae jump onto a fly and, if not removed by grooming, feed, and remain there until the fly dies. Nymphs and adults of the mite feed on brine fly eggs: "the pattern in space of resources for both parasitic and free-living stages of the mite will be a mosaic of high and low density patches with few patches of intermediate density" (Collins, 1975:250). Costs due to low probability of colonization are high: "the 76%-91% mortality associated with *Partnuniella*'s parasitic stage may be an irreducible cost of exploiting an unpredictable and extremely clumped distribution of hosts, combined with host defenses which effectively filter out adaptations that could improve hostfinding ability" (Collins, Mitchell, and Wiegert, 1976:1229).

Heteroecism in parasites of plants and animals (as in

Schistosoma) undoubtedly increase the patchiness of niche space since the two or more hosts must overlap or abut in their niche occupation before a resource patch is created for the parasite. This overlap is likely to be particulary narrow for heteroecious parasites of plants since usually one host species is woody and perennial and the other an herbaceous annual. The hosts occupy different stages in ecological succession. As an example, the ecology of the rust fungus *Puccinia graminis* may be scrutinized (see Figures 3.5 and 3.6 for summary).

The organisms involved in black stem rust epidemiology include the pathogen *Puccinia graminis,* summer hosts in the Graminae (wheat, *Triticum*; oats, *Avena*; barley, *Hordeum*; and others), a spring host barberry *Berberis vulgaris,* and insects that transfer pycniospores from one pycnium to

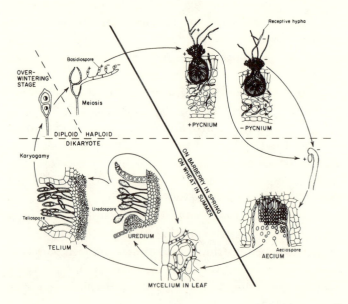

FIGURE 3.5. Life cycle of *Puccinia graminis* (after Alexopoulos, 1952)

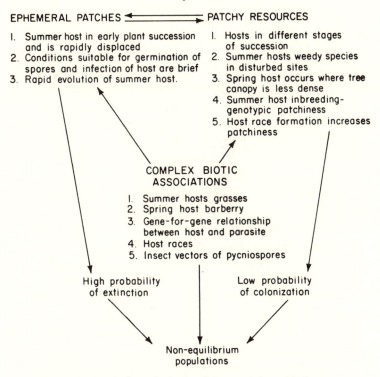

EPHEMERAL PATCHES ⟷ PATCHY RESOURCES

1. Summer host in early plant succession and is rapidly displaced
2. Conditions suitable for germination of spores and infection of host are brief
3. Rapid evolution of summer host.

1. Hosts in different stages of succession
2. Summer hosts weedy species in disturbed sites
3. Spring host occurs where tree canopy is less dense
4. Summer host inbreeding-genotypic patchiness
5. Host race formation increases patchiness

COMPLEX BIOTIC ASSOCIATIONS

1. Summer hosts grasses
2. Spring host barberry
3. Gene-for-gene relationship between host and parasite
4. Host races
5. Insect vectors of pycniospores

High probability of extinction

Low probability of colonization

Non-equilibrium populations

FIGURE 3.6. Patch dynamics in *Puccinia graminis* populations.

another. The summer hosts in their primeval state were mostly annual weedy species occuring in disturbed habitats in very early stages in plant succession. *Berberis vulgaris* is a woody perennial found in a much later stage of succession in woodland vegetation, particularly where the canopy is less dense. The juxtaposition of the two host species is critical to survival of the fungus in areas where summer spores cannot survive the winter; it also limits in space and time the potential resources for *P. graminis*. The summer hosts particularly will be highly patchy in distribution being dependent upon disturbance, and only those patches close to barberry could be colonized by the fungus. Moreover, the grass species such as *Triticum* disperse seeds poorly in space but more effectively in time; the seeds lie dormant on the site of the parents until another soil disturbance makes conditions for germination suitable (see Harper, 1977). *Triticum* is also a selfing species (Large, 1940), so small patches are likely to contain individuals homozygous at many loci and each patch is likely to differ genetically in many traits. The potential for rapid evolution is present. Host race formation is well developed.

Resistance factors such as hypersensitivity of cells, the level of production of the antibiotic DIMBOA, and percentage penetration (see Deverall, 1977) will differ radically between patches. There are 20 identified single genes for resistance to black stem rust on wheat (Day, 1974) that could result in 2^{20} combinations of genes in natural populations, and therefore 2^{20} races of the fungus to exploit this variety of hosts fully. As Day (1974) points out, the gene-for-gene relationship between host and fungus has been demonstrated repeatedly for the rust on wheat (Green, 1964, 1966; Kao and Knott, 1969; Loegering and Powers, 1962; Williams, Gough, and Rondon, 1966; Luig and Watson, 1961) and on oats (Martens, McKenzie, and Green, 1970). Not only will the host species be patchily distributed, but for any strain of fungus the susceptible strains of host will be even more patchy. Since the rust can frequently reduce seed set to zero, the dynamics of any genetic strain of host will usually be in a state of flux. But persistence of a host strain in time will be increased by the presence of the same genotype in the seed bank.

The barberry host also differs considerably in resistance to infection from basidiospores according to species and age of leaves (Melander and Craigie, 1927). Those species with the thickest outer epidermal wall and cuticle on young

62

leaves, such as *Berberis thunbergii*, were most resistant. But all species showed increasing thickness of wall and cuticle with increased age of the leaf rendering a plant most vulnerable at the first flush of foliage in the spring.

The life cycle of *Puccinia graminis* obviously has two weak links involving the colonization of a summer and a spring host. Aeciospores (dikaryotic) are produced in aecia on the undersides of barberry leaves in the spring. They are blown on the wind to grasses, penetrate the leaves, and soon give rise to uredospores (dikaryotic) that infest members of the same summer-host species. As autumn draws near, dark brown, thick-walled, two-celled teliospores (dikaryotic) are formed on grasses; they fall to the ground and overwinter. On germination in the spring each cell of the teliospore gives rise to four basidiospores (uninucleate). These spores are carried on the wind, and some must reach barberry plants. Here they germinate, ramify the *Berberis* leaf, and eventually produce pycnia (spermagonia) in which pycniospores (uninucleate) are produced. One pycniospore fuses with a receptive hypha of another pycnium, and soon aecia containing aeciospores (dikaryotic) are formed. The cycle is completed.

Arrival of an aeciospore on an appropriate host of both the right species and the right genotype, at the right stage in leaf development, and in the correct microclimate is a tremendous colonization problem in itself and produces a weak link in the life cycle. But the eventual production of aeciospores is preceded by an even more tenuous series of events, only alleviated to some extent by those age old associates of parasites: the insects. *Puccinia graminis* is a heterothallic species: basidiospores germinate to produce a thallus with nuclei of a positive or negative mating type. The dikaryotic thallus that produces aeciospores is produced by fusion of a positive pycniospore with a receptive hypha produced by a negative pycnium, or vice versa. Thus for the cycle to be completed two avenues are open. Two basidiospores of opposite mating type must be blown a considerable distance to the barberry, and they must germinate on and ramify in a single leaf (so that a pycniospore of one mating type can be readily transferred to the receptive hypha of the opposite mating type) enormously reducing the probability of successful colonization compared to the need for only one spore for colonization. The other alternative is that basidiospores reach and germinate on different leaves or even different plants and insects carry pycniospores from one plant to another. The pycnia produce a nectar-like mucus, and a mixture of this and the spores is exuded onto the upper surface of the leaf. Many insects such as flies, sawflies, and wasps forage on leaf surfaces for sugary secretions, and their high mobility and directed flight greatly increase the probability of union between mating types. Nevertheless, however omnipresent flies and other insects seem to be, they add biotic complexity to the life cycle of *Puccinia graminis* and greatly enhance the multiplicity of factors impinging on its patch dynamics.

Where, then, are the forces in the *Puccinia graminis* system that generate equilibrium? In my estimation they are trivial in the face of factors contributing to non-equilibrium within patches (see Figure 3.6).

NON-EQUILIBRIUM COMMUNITIES

Parasite communities may be viewed either as the coexisting species on or in a single host if the host is relatively large compared to the parasites like in the gut communities of vertebrates or as coexisting species on a host population or patch, an approach normally adopted in studies of parasitic herbivores. But in both cases, with each parasite spe-

cies existing in non-equilibrium conditions, it is most unlikely that the resulting guild or community will be in
equilibrium. At the community level equilibrium is defined
as the state in which the rate of immigration of new species
is equal to the rate of extinction of resident species (Mac-
Arthur and Wilson, 1967). The equilibrium theory of island
biogeography predicts that for patches of similar size and
disposition the equilibrium number of species should be
similar and stable if patches remain undisturbed. These
properties should be included in the definition. A patch
with equal rates of immigration and extinction need not
be in equilibrium, for the rates may change with time
and an equivalent patch may harbor a different number
of species.

The mechanisms that foster non-equilibrium community
states are the same as those for non-equilibrium populations depicted in Figure 3.2. If for any host or host patch
colonization has a low probability and extinction has a high
probability, resources will usually remain unsaturated.
Any parasite species may become extinct independently of
the presence of others, and extinction may occur even if
only one species is present. This contrasts with the equilibrium state in which the probability of extinction increases
with "increasing probability of interference among species"
that "will have an accelerating detrimental effect" (Mac-
Arthur and Wilson, 1967:22).

Some examples may clarify the dynamic nature of species
presence and absence, and introduce influences not discussed under non-equilibrium populations. These include
interaction between parasite species and coexistence, the
effects of differential colonizing ability, and the response
of parasite species to habitat complexity.

For the community of parasitic wasps that exploit the
Swaine jack pine sawfly *Neodiprion swainei*, resources and

the realized niches of the wasps are in a dynamic state in space and time (Price, 1973b) due to several interacting factors.

(1) The host sawfly has complete metamorphosis and one generation per year, much of which is passed in a cocoon during cold weather when parasites are inactive. The egg, free-living larva, and cocooned larva and pupa offer very different resources, and parasites have specialized to exploit only one stage. Each stage is available for a relatively short time in each generation making colonization and exploitation of a host population less effective than on stable resources.

(2) Particularly in the larval stage, the host is changing so rapidly in abundance, size, and dispersion that any one parasite is not able to exploit the full range of larval instars: each species has specialized on young, or mature larvae, or larvae just about to spin a cocoon (Price, 1974a). Young larvae are very small, colonial, and relatively abundant. The survivorship curve shows high mortality through the larval stage (McLeod, 1972); colony size decreases until the final instar larvae become solitary, drop to the ground, and spin cocoons in the forest litter. Again utilization of such evanescent resources, with shortage of time dominating the scene, results in the underexploitation of hosts. Andrewartha and Birch (1954) emphasized similar prevailing influences for organisms in general, although their view was largely influenced by the study of parasitic insects.

(3) Jack pine *Pinus banksiana* is patchily distributed, being found mostly on sandy outwashed plains, and it regenerates luxuriantly after fire to be succeeded with time by black spruce and hardwoods. Small stands of jack pine also exist in other dry sites. The sawfly, particularly at the beginning

of an epidemic, occupies relatively small areas within these jack pine stands, the epicenters of McLeod (1972). Within each jack pine stand host cocoons must be searched for in a complex matrix of litter types, patchily distributed from very dry open sand to lichen cover, moss cover, and wet hardwood litter. Parasites specialize in their niche exploitation on this moisture gradient, so patchiness on this microgeographic scale is a reality for them.

(4) Within each patch the sawfly population is in a state of flux, increasing and decreasing in numbers per unit area, and in the area occupied, as more marginal sites are colonized with increasing density at the epicenter. Parasites are exposed to rapidly increasing resources over a 3 or 4 year period and then a catastrophic decline in 1 or 2 years that compresses the parasite populations into a brief period of intense competition.

(5) Some parasite species that utilized host larvae are adapted to colonize early in a host epidemic, but they are poor competitors. Others disperse slowly, colonize late, but once established overcome the early colonizers. Tenure of a site may be abbreviated by competitive interaction. Even within the slow colonizing guild, members of which attack the cocoon stage, some exploit low host populations better than others and are again overcome as host populations increase.

The dynamic properties of the parasite community of *N. swainei* may be summarized for time and space. As host density waxes and wanes at a site, the species diversity of parasites is also in a state of flux (Figure 3.7). Rapid colonizers from the larval parasite guild arrived first contributing 100 percent of the species diversity (H') at the two densities sampled below 1 host per 30 centimeters square.

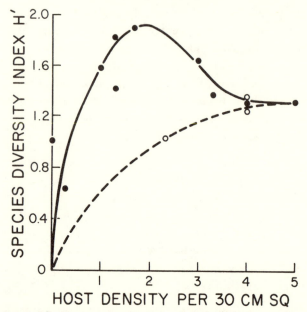

FIGURE 3.7. Change in species diversity of parasitic Hymenoptera in increasing host populations represented by closed circles (·) and solid line and decreasing host density represented by open circles (o) and dashed line indicating hysteresis in the system (after Price, 1973a)

Then cocoon parasites arrived, after which time the larval parasites did not contribute more than 40 percent to total H' in any sample. As two cocoon parasites became dominant in the community, other species were pushed to extinction, and beyond 3 hosts per 30 centimeters square one species commanded a majority of the field and maintained this position even in declining host populations. Thus parasite communities exhibited lower diversity in declining host populations than in increasing host populations, a hysteresis in the system that lessens the probability of any community equilibrium being observed. A similar pattern may be seen in space (Figure 3.8). At the edge of a host population,

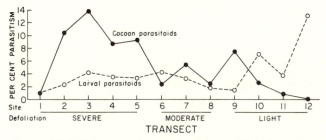

FIGURE 3.8. Relative abundances of cocoon and larval parasites from the edge of a large host population (Site 12) to the center (Site 3). (Site 1 was near the epicenter of the host population but adjacent to a lake.) (After Price, 1973b).

equivalent to the beginnings of an epidemic, larval parasites cause all the parasitism; this dominance declines as host density increases and cocoon parasites become more influential. The low levels of parasitism illustrate the apparent superfluity of resources to be expected in non-equilibrium communities. Time is the limiting resource rather than energy or nutrients.

Communities of parasitic herbivores on annual and biennial plants also exploit a host population dynamic in space and time. Harper (1977) emphasizes the seasonality of vegetative growth. As in the wild parsnip example given in the non-equilibrium population section of this chapter, resources are constantly changing. Seedlings offer little protection and little niche diversity to parasites. As the season progresses both elements increase, and, with senescence, they decline again. The microclimate is greatly ameliorated for pathogens where the gross production of plants is high (e.g. Hirst, 1958). In such a system the number of species present in the community changes constantly. An extreme case of an agricultural crop was studied by Price (1976) and Mayse and Price (1978) and a more general case using the aerial parts of bracken, by Lawton (1978). In the latter case

69

Lawton (1976, 1978) also illustrated the dramatic change in plant quality throughout the season involving allelochemicals and nutrients. Even for small herbivores on trees the most convincing hypothesis to account for coexistence of eight species of leafhoppers invoked non-equilibrium conditions (McClure and Price, 1976). Wilson (1951) found that the number of fungi and bacteria in the rhizosphere of beech changed constantly through the season, just as the herbivores on soybean and bracken, whereas the number of free-living species in the soil was much more stable.

Extensive studies by Dogiel (1961) on the parasites of fish led him to conclude that the diversity of parasites increase with fish age, and we must conclude that no equilibrium number was reached in the majority of cases. Kennedy (1975a) also found in the River Exe that the longer young salmon remained in the river the more diverse the parasitic fauna became. Living organisms frequently provide islands too short-lived for community equilibrium to develop.

There is generally a poor relationship between vegetative diversity and parasitic herbivore diversity (for details and discussions see Kostrowicki, 1969; Murdoch, Evans, and Peterson, 1972; Sharp, Parks, and Ehrlich, 1974; Mellinger and McNaughton, 1975; Rathcke, 1976; Gibson, 1976; Slansky, 1976). This should be expected where probability of colonizing each patch is low and the probability of local extinction is high for each species of parasite in the community. Under non-equilibrium conditions resources will normally be underexploited and species underrepresented.

Long-term studies on parasites are rare and so we have little idea of the probabilities of colonization and extinction for parasites. The studies of Slapton Ley in England over the past decade give us cause to be concerned with the importance of rare events (see Kennedy, 1975b for details). In

1973 a pair of great crested grebes bred on the Ley for the first time in many years. That was the first year in many in which a serious pathogenic cestode *Ligula intestinalis* was discovered in the lake (in one fish), almost certainly introduced by the grebes. Since its introduction *Ligula* has increased dramatically: 5 fish were infected in 1974 and 40 percent of the roach in 1975. Kennedy anticipates an epidemic. *Ligula* attacks many freshwater fish species (Kennedy, 1974), but is most commonly found on roach in which it causes castration (Arme and Owen, 1968). Therefore the parasite will probably depress the roach population, releasing other species from competition. The fish community may change profoundly, with roach possibly being pushed to extinction, accompanied perhaps by the demise of *Ligula intestinalis*. The value of Kennedy's long term study is that the real impact of this rare colonization event will be followed to its conclusion. In addition the non-equilibrium state in the *Ligula* population (mimicking the population in Figure 3.1b so far) and the importance of chance in colonization for parasite and potential hosts are well illustrated.

Another case with a shorter cycle of colonization and extinction investigated by Morris (1970), concerns the aquatic snail host *Helisoma trivolvis*, its trematode parasite *Echinoparyphium recurvatum*, and the definitive host that in this case was blue-winged teal *Anas discors*. The study was stimulated by the observation that snails were found in only a small fraction of the many apparently suitable ponds available in central Alberta. Events probably take the following course. A pond is colonized by the snail that undergoes a population flush. Teal are attracted to an abundance of snails and, while feeding, introduce eggs of the trematode. A population explosion of the parasite follows causing castration of infected adult snails and excessive mortality of juveniles. The snail population crashes to very low levels or to extinction. The parasite follows suit. The rapidity of the cycle relative to the probability of colonization by snails leaves many sites unoccupied.

Many of the aspects of the interaction between parasites and hosts at the population and community levels discussed in this chapter have been integrated in studies by Sankurathri and Holmes (1976a, b): the intricacies of biotic relationships, their highly dynamic state, the role of mutualisms and lack of coexistence under certain conditions, the importance of external factors and their differential direct and indirect influences on all biotic components of the system (Figure 3.9). The study involved the aquatic snail

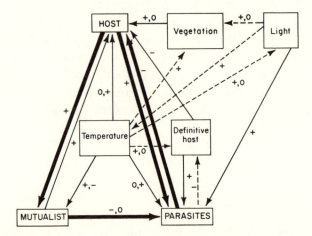

FIGURE 3.9. Interactions between the snail host, its digenean parasites, an oligochaet mutualist, and physical factors as discussed in the text. Major biotic relationships included in the study are shown by heavy lines and arrows. Other solid lines indicate direct effects; dashed lines indicate indirect effects. The influences are shown as positive (+), negative (−), a combination of both (+, −), or a combination of positive and zero effects if influence changes with the season (+, 0 and 0, +) (after Sankurathri and Holmes, 1976b).

Physa gyrina, a mutualistic oligochaet worm *Chaetogaster limnaei*, the community of digenean trematode parasites of the snail, numerous definitive hosts, aquatic plants as food for the snail, and abiotic influences on these organisms.

Biotic interactions included an increase in snail populations with an increase in aquatic plants. As snail populations increased, so the symbiotic oligocheat that lives in the mantle cavity of *P. gyrina* increased. But with more oligochaets the numbers of digenean parasites infecting hosts declined because the oligochaets actively fed upon miracidia and cercariae and physically blocked access to the snail host. However, water temperatures in the summer became excessive for the worm and populations crashed; infection by parasites increased causing a decline in snail populations by reducing the fecundity and longevity of hosts.

The abiotic influences were principally light and temperature. Light affects miracidial behavior and the growth rate of aquatic plants. Temperature affects: (1) growth of aquatic plants and snails; (2) ice formation that limits access to water by definitive hosts and light intensity in the water; (3) reproductive activity of parasites; (4) growth and mortality of oligocheats. These influences change dramatically as thresholds are reached and passed. For example as temperature increases, oligochaets are initially favored until temperature becomes excessive in August (indicated by +, − in Figure 3.9). In other cases temperature may have no effect during part of the season and a positive affect at other times (o, + or +, o in Figure 3.9). One is left with the clear impression that no resource in this system remains accessible for long enough to permit the development of equilibria in the parasite populations and communities.

EQUILIBRIUM CONDITIONS

In situations where the probability of colonization by parasites is high, it may seem that communities should reach equilibrium conditions. Particularly in aquatic environments colonization should be easy, and, indeed, high infection rates can be found, for example, in acanthocephalans in salmonid fish (Holmes, Hobbs, and Leong, 1977) and *Schistosoma* in humans (Hairston, 1965; Cohen, 1976) and snails (Berrie, 1973), but by no means do all host species have a high rate of infection. However, after careful review of the literature, Kennedy (1977) concluded that the majority of fish parasite populations exist in non-equilibrium conditions, a generalization that also seemed valid for parasites on the majority of poikilotherms (Kennedy, 1975a). Even where parasite-host population models have predicted an equilibrium state, such as those of Crofton (1971) and Anderson (1974), populations used were large (10^3-10^6) and fluctuations frequently were greater than one order of magnitude. If such fluctuations occurred in small populations, extinction would commonly result.

The considerable literature on niche differentiation in coexisting parasites in fish may also lead to the conclusion that competition is common and species packing is tight (e.g. see Sogandares-Bernal, 1959; Chappell, 1969; MacKenzie and Gibson, 1970; Williams, McVicar, and Ralph, 1970; Holmes, 1971, 1973; Hair and Holmes, 1975). But as argued earlier in this chapter, competition may be intermittent. For example, after a host population crash, the parasite fauna is compressed onto a relatively small number of hosts. Kennedy (1977) found no evidence that interspecific competition played an important role in fish parasite population regulation. An extreme and unusual case of tight species packing among pinworms in the alimentary canal of tortoises studied by Petter (1962, 1966) and Schad (1963a, b) indicates high probabilities of colonization. Just why infection by pinworms of terrestrial hosts is so effective is not clear, but there is no intermediate host and infection can occur by inhaling eggs or by consuming contaminated food or water. Clearly the subject of species packing in parasite communities needs careful examination, and this will be undertaken in Chapter 6.

It should be remembered that the complexity generating patchiness may be unimportant to the epidemiologist who is interested principally in the mean of patches. As Cohen (1976) notes, disease ecology may seem to involve unmanageable complexity but simple models in schistosomiasis have been very successful.

For the parasite, also, much of the environmental complexity is filtered out because of sensitivity only to highly specific cues or to cues in time that synchronize parasite activity with host availability. But the patchiness of resources for parasites in space and time is nevertheless real, and the probability of existence in a non-equilbrium state within patches is high. The adaptive syndromes to cope with this milieu are the subject of the next chapter.

CHAPTER FOUR

Genetic Systems

The union between population ecology and population genetics has not been achieved in spite of recognition by many that the two aspects of population biology are inextricable. As organisms and effective population sizes become smaller and/or less mobile, so recognition by biologists of the importance of interaction among genetics, cytology, and population size and structure seems to increase. In parasite-host systems complex interactions may exist in both host populations and parasite populations. Of snails in the genus *Bulinus*, which act as hosts to *Schistosoma haematobium*, Berrie (1973:176) stated "Ecological and genetic factors are constantly interacting so that individual populations may develop quite distinctive features." Of parasitic mites on plants in glasshouses Helle and Pieterse (1965:308) said, "There is a patchwork quilt of genetic incompatibilities between closely adjacent populations." Harper (1977:412) remarked concerning phytophagous parasites in general, "The population dynamics is continually confounded in the population genetics."

The only logical approach to the study of population ecology or population genetics is to unite them both into a study of the genetic systems of organisms. In this book because of its emphasis on evolutionary biology, the levels of selection (individuals and populations) and the units of evolution (populations and species) must be given a central place. The population (the deme or local population of Mayr, 1963) is defined for bisexual species as the potentially interbreeding individuals in a locality. White

(1954) defines the genetic system as all characteristics determining the hereditary behavior of a species. These characteristics include the mode of reproduction, the chromosome cycle, breeding system, population structure (including polymorphism) and dynamics. They define the "kind of species" in the sense of Mayr (1963). He listed 13 criteria that may be used to classify species biologically into kinds: system of reproduction, degree of intra- and inter-specific fertility, presence or absence of hybridization, variation in chromosome number and pattern, difference in origin, structure of species, size of populations, sequence of generations, amount of gene flow, pattern of distribution, environmental tolerance, rate of evolution, and phenotypic plasticity. Mayr's list will be used to summarize differences in kind of species between predators and parasites at the end of this chapter. These factors combine to determine the species' evolutionary potential, a critical aspect of the species' adaptive syndrome (i.e. the major intercorrelated adaptive features of a species; Root, 1975; Root and Chaplin, 1976).

The ecological setting for a population provides the independent variables in the genetic system. Resource structure in relation to size and the mobility and tolerance of individuals must be known before the adaptive qualities of intercorrelated properties of the species can be evaluated. The patchiness and ephemeral nature of resources for parasites were emphasized in Chapter 3. The properties of parasitic species will be examined in relation to this ecological background and to the consequent low probability of colonization and high probability of extinction in patches.

The intercorrelated properties of species that adapt them to exploit such resources can be divided into three major subjects: breeding system, mode of reproduction, and population structure. Underlying these factors will be the consideration of how genomes and karyotypes change through such influences, that is, how the populations and species evolve, and the modes of speciation. A synthesis of parasite population ecology and population genetics, although desirable, can not be presented in a small chapter, even if the relevant data were available. Therefore, emphasis will be placed on the adaptive syndromes of parasites.

A result of the low probability of patch colonization is that when an individual arrives at a patch it is likely to find no others there. None may arrive during the reproductive life of the colonizer. In addition, an ephemeral patch, once discovered, must be exploited rapidly before it becomes untenable. The surest way of overcoming these ecological strictures on the breeding system is to mate before dispersal, to be hermaphroditic and self-compatible, or to be parthenogenetic and not to mate at all. These strategies are commonly observed in parasitic organisms.

Mating before dispersal is seen in many groups of parasitic organisms with mobile adults. Thysanoptera usually mate before leaving the host plant, and, understandably, some species have wingless males (Lewis, 1973). The tetranychid mites that disperse from host plants on silken threads do so after mating (Mitchell, 1970). Mites parasitic on the eggs of scolytid beetles reproduce on a single host egg walled off from others (Mitchell, 1970; Lindquist, 1969). Sixty progeny may be produced of which only three may be males. Before feeding, the males mate with siblings and die within the cell. Inseminated females escape before feeding, climb onto newly emerged adult beetles and by phoresy reach new breeding sites where they feed and mature. The potency of the selective force to mate early is seen here where mating preceeds even feeding. In the moth-ear mite *Dicrocheles phalaenodectes* males feed before mating, but they develop through fewer stages than females;

78

they are most likely to mate with siblings, and do not leave the inner parts of the ear (Treat, 1975). It is among the parasitic insects that we see examples of apparent rape; copulation that occurs before the female has escaped from her pupal position and before she is able to exercise any choice. Three or four males of *Limothrips denticornis* (Thysanoptera) mate with about twenty-five female pre-pupae (Lewis, 1973) rather like the mite example above. Gilbert (1975) has found that males of the butterfly *Heliconius charitonia* routinely rape the female pupa. Copulation is also attempted by males of *Megarhyssa* spp. immediately after a female has emerged from her pupal chamber leaving her no choice in the union (Heatwole, Davis, and Wenner, 1964). From the females' point of view this may indicate that the males most likely to mate with a female have similar genotypes and thus choice among them is unimportant. In all cases where mating occurs before dispersal, sib mating will be common and probably prevalent. Inbreeding such as this has far-reaching consequences for population structure and evolutionary potential, which will be discussed later in this chapter.

A major alternative to mating before dispersal is for mates to meet at an easily recognized site. Such sites include trees and fruits for fruit flies (Prokopy, 1968; Moericke et al., 1975) or hilltops for many flies (Diptera) and butterflies (Lepidoptera) (Shields, 1967). Both flies and butterflies are highly mobile and visually oriented in contrast to many parasites that perceive largely chemical cues and have poor means of locomotion. Such breeding systems result in outbreeding populations and the maintenance of relatively large effective populations. While resources may be very patchy and isolated, adults from every patch in the vicinity congregate at a hilltop to mate, and then females colonize new patches. Many families of parasitic flies are known to congregate on hilltops (Shields, 1967): Agromyzidae, Anthomyiidae, Bombyliidae, Calliphoridae, Cuterebridae, Gasterophilidae, Oestridae, Phoridae, Sarcophagidae, and Tachinidae. By contrast, only the Ichnemonidae among other families of parasitic insects are cited a few times as "hilltoppers." It is tempting to find in this major difference in breeding system part of the reason for the lower species richness of parasitic flies than parasitic wasps in any fauna. For example, in Britain parasitic flies represent 6 percent of the fauna whereas parasitic wasps represent about 28 percent (cf. Table 1.1).

Many parasites disperse to new hosts in immature stages, so mating before dispersal is not available as a strategy. This greatly reduces the chances of finding a mate, and there is a strong selective pressure for doing without. It is very instructive to look at the primitive facultative parasitic nematodes in this regard. *Rhabdias bufonis* has a free-living generation of both males and females, but the parasitic generation in the lungs of frogs and toads is composed of hermaphrodites (Baer, 1971). When mates are unavailable, self-fertilization is possible. *Strongyloides stercoralis*, which causes strongyloidiasis in primates and other vertebrates, reproduces in the soil through amphimixis while conditions are favorable (see Table 4.1 for definition of reproductive modes). When conditions deteriorate, an infective generation is produced from matings of free-living male and female nematodes that mature in a host and reproduce parthenogenetically (Noble and Noble, 1976). Two closely related plant parasitic nematode genera are *Heterodera*, which is mostly amphimictic, and *Meloidogyne*, which is mostly parthenogenetic (Franklin, 1971; Triantaphyllou, 1971). *Heterodera* females mate outside of the host plant with free-living males whereas *Meloidogyne* females are complete endoparasites living in root galls

TABLE 4.1. Classification of modes of reproduction in soil-dwelling and plant parasitic nematodes (data from Triantaphyllou, 1971). Modes are arranged in an order showing a degradation from sexual reproduction resulting in a trend of decreasing mean variability among progeny from a number of individuals reproducing. The percentages of non-parasitic and parasitic nematodes in each reproductive mode are given.

Mode of reproduction	Percentage of non-parasitic species	Percentage of parasitic species
Amphimixis[1]	38	49
Amphimixis[1] plus automixis[2]	10	0
Amphimixis[1] plus pseudogamy[3]	5	0
Amphimixis[1] plus parthenogenesis (meiotic)[4]	0	11
Automixis[2]	19	0
Pseudogamy[3]	14	0
Parthenogenesis (meiotic)[4]	5	7
Parthenogenesis (mitotic)[5]	10	33
TOTAL SPECIES	21	45

[1] Fertilization of oocyte by sperm from another individual.

[2] Fertilization of oocyte by sperm from the same individual (hermaphrodites).

[3] Activation of oocyte by sperm penetration but with no fusion of nuclei.

[4] Production of a haploid oocyte through meiosis which then fuses with a nearby haploid cell (meiotic thelytoky).

[5] Production of a diploid oocyte through mitosis (mitotic thelytoky).

(Franklin, 1971). In a survey of modes of reproduction in soil-dwelling and plant parasitic nematodes, Triantaphyllou (1971) considered data on available species that demonstrated all stages in the progression from amphimixis to mitotic parthenogenesis showing gradual decay of sexual reproduction: the progression ranged from maximum variability among progeny from a mating to no mating and virtually no variability among progeny (Table 4.1). From the distribution of numbers of species in each reproductive mode it appears that the major adaptive modes for parasitic nematodes are the extremes of pure amphimixis at one end and mitotic parthenogenesis at the other but with another significant peak involving amphimixis plus parthenogenesis. Evidently evolution to mitotic parthenogenesis can proceed rapidly leaving relatively few species at intermediate states at any one time. It is significant that many parasitic nematodes are found toward the parthenogenesis end of the trend.

The commonness of parthenogenesis among parasitic organisms has already been alluded to in Ecological Concept 1 (Chapter 2). If we can understand the adaptive features of this reproductive mode, which in mitotic parthenogenesis ensures duplication of the materal genotype (except for mutation), we can probably understand all the other modes. This assumes acceptance of William's (1975) view that sexual reproduction leads to greater variance in fitness dosage among progeny with maximum fitness significantly greater than among asexual progeny (Figure 4.1). The modes of reproduction intermediate between amphimixis and parthenogenesis are either best-of-both-worlds strategies (life cycles involving both sexual and asexual reproduction) or non-equilibrium states between the two extremes. For example, pseudogamy, where a spermatozoan stimulates the egg to develop parthenogenetically, must surely be a transitional state on the way to parthenogenesis without the need for spermatozoa.

The adaptive features of parthenogenesis are numerous when short term survival of a clone is considered.

(1) A single female can establish a new colony (White, 1973; Williams, 1975; Maynard Smith, 1978), so multiplication can occur when the probability of contact between conspecifics is very low (Tomlinson, 1966).

(2) The gross reproductive rate of parthenogenetic females is virtually doubled (White, 1973; Williams, 1975), so the probability of finding a new host by progeny is similarly increased.

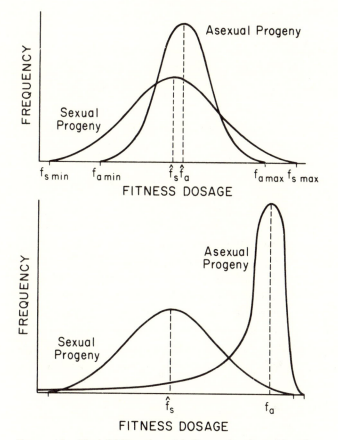

FIGURE 4.1. (Top) Williams' model illustrating favorable selection for sexuality through greater variation in fitness with consequent achievement of higher fitness among some progeny (cf. $f_{s\ max}$ and $f_{a\ max}$), although mean fitness dosage (f_s) is lower than for asexual progeny (f_a) (after Williams, 1975). (Bottom). For asexual parasites, and particularly those reproducing by mitotic parthenogenesis, most progeny will have a genotype very close to the adaptive peak for the clone because the maternal genotype is already closely adapted to host resources. A small proportion carrying new favorable mutations may reach beyond the fitness that can be achieved by sexual reproduction because recombinational load is zero. The tail of the distribution represents progeny carrying new unfavorable mutations. The comparison is made for asexual species utilizing highly predictable resources and sexual species exploiting typically unpredictable resources.

(3) Where close inbreeding is very common, parthenogenesis provides results very similar to those achieved by sexual reproduction. There is also a greater probability of parthenogenesis evolving in inbred populations (Cuellar, 1977).

(4) The variance of fitness among progeny is low compared to that in sexually produced progeny, and fitness is clustered close to the adaptive peak of the clone. The majority of progeny have high fitness once established in a host, and mean fitness among progeny will be higher than in sexually produced progeny that carry a recombinational load (Williams, 1975).

(5) Parthenogenetic parasites appear to be maladapted to a patchy, ephemeral environment since variability among a female's progeny will be small (Mayr, 1963; White, 1970; Williams, 1975). However, this disadvantage can be reduced by utilization of highly stable and predictable microenvironments provided by the homeostasis of living organisms. Glesener and Tilman (1978) also suggest that parthenogenesis will be favored in environments with high biotic and abiotic predictability. A positive feedback loop will sustain the parthenogenetic mode (Figure 4.2).

(6) Particularly adaptive combinations of genes are fixed (except for mutation) without danger of disruption (see also John, 1976). No recombination occurs in mitotic parthenogenesis, and in meiotic parthenogenesis crossing-over occurs between homozygous chromosomes. This locking of gene combinations may be especially important in parasitic organisms where large banks of genes are likely to be involved with close coevolutionary tracking of the host system. Disruption of such a block would generate gross maladaptations with almost certain lethal results. Another positive response will reinforce the parthenogenetic mode (Figure 4.2).

84

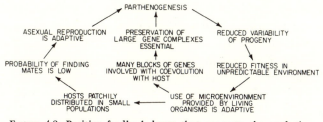

FIGURE 4.2. Positive feedback loops that promote the evolution of parthenogenesis among parasitic organisms (from Price, 1977).

In the long term parthenogenesis has usually been regarded as an evolutionary dead end, as a short-term strategy doomed to eventual failure (e.g. Maynard Smith, 1978; White, 1978). But a reappraisal is warranted on several counts.

(1) Considerable variability has been found in parthenogenetic species (e.g. Suomalainen and Saura, 1973; Suomalainen, Saura, and Lokki, 1976; Atchley, 1977) providing ample material on which natural selection can act.

(2) In many instances the host genotype changes relatively slowly because of a long generation time compared to that of the parasite. In an extreme case of a multivoltine parasite on a long-lived tree, the parasite may pass through 1000 generations during the life of a single host. Then this tree is likely to be replaced by a close relative. In ephemeral plants genotypic continuity at one site is provided by the seed bank, so parasites with long-lived resting stages can track this resource, although the ratio of generations will be much closer. Where generation times differ considerably, there must be a strong selection to *slow down* the evolutionary potential of the parasite either through parthenogenesis or by other means that reduce recombination (see also Bonner, 1965; Sonneborn, 1957).

(3) Massive eruptions and dramatic crashes in population size are likely to be common, as stressed for non-equilibrium populations (Chapter 3). Stebbins (1958) and Harper (1977) have defined the breeding system best adapted to circumvent this genetically disruptive situation. In outbreeding organisms a population flush after relaxed selection liberates an enormous array of genetic variability with most genotypes unfit for the intervening periods. Harper (1977:773) describes the situation as follows: "After each explosion the population returns to the normal condition less fit, unless in some way the display of variation has been controlled during the colonizing phase. We may perhaps interpret the predominantly inbreeding habit of colonizers and fugitives as a protection against genetic display during colonizing episodes and against the too rapid tracking of adaptive optima that are only transient. . . . Close inbreeding or apomixis are ways in which temporary forces of selection are prevented from throwing the long-term adaptation of a population off course."

(4) Judging from the results of breeding resistant varieties of agricultural plants and the resulting gene-for-gene concept, we see that the evolutionary potential of a parasite species or clone depends more on new genes than it does on reassortment of genes (see also Bush's model on the genetics of host shifts as discussed on pp. 98-101). Parthenogenesis in no way lessens mutation rates as far as we know, and the expression of mutant genes is not severely altered except in polyploid clones.

(5) Short generation time (see also 8 below) and high fecundity result in a large number of individuals being produced per unit time relative to hosts or free-living organisms of a similar size. The number of mutations in a clone will also be relatively high and will probably provide

enough variation to permit the long-term viability of a clone.

(6) Perhaps for small, highly specialized organisms with narrow tolerances, sites with conditions very similar to the maternal environment can be reached even after a colonizing event. There may be less of a premium on major genotypic reconstructions for such specific organisms than in free-living species that seem to be exposed to a more variable environment. Indeed, Askew (1975) found that the more specific were parasite species attacking the contents of oak galls, the higher was the ratio of females to males emerging, until only when parasite species attacked one or two hosts did some thelytokous species occur. Where hosts were predictably similar, parthenogenesis could evolve. Conversely, where hosts attacked are very diverse and almost totally unpredictable, as in angiosperm parasites such as *Orthocarpus* and *Euphrasia*, obligate outcrossing is usual and gene flow is even interspecific, contributing significantly to the genetic variability of any one species (Atsatt, 1970; Yeo, 1978).

(7) Where fecundity is very high and colonization is hazardous, and in many cases completely stochastic, survival through the most difficult stage in the life cycle does not depend on genotype but on chance. Again the conservative strategy would seem to maximize parental fitness, for if progeny have a narrow range of fitness, any one is likely to perpetuate the clone once colonization is achieved.

(8) Parthenogenesis facilitates the evolution of progenesis (Gould, 1977). Gould defines progenesis as the retention of formerly juvenile characters by adult descendents produced by precocious sexual maturation of an organism still in a morphologically juvenile state. Parthenogenetic species can

87

readily adapt through progenesis because the many adult characters associated with mating and copulation are unnecessary. Juvenile forms can reproduce effectively. The adaptive feature of progenesis is that rapid maturation is possible, and therefore the intrinsic rate of natural increase can be very high. Gould (1977) identifies three conditions under which progenesis is highly adaptive: (1) unstable environments; (2) where colonization is frequent; (3) where small size is adaptive. All these features are common to parasites, a fourth category that Gould recognizes. Wright (1971) describes some extreme examples in flukes (Trematoda: Digenea).

These points demand a reevaluation of the adaptive nature of asexuality among parasites as illustrated in Figure 4.1.

The cost of evolving may become a major feature in the population dynamics of a species, a cost that may be best minimized in many parasites by parthenogenesis. Morris (1969) describes many ways in which temperature enters into his population model on the univoltine fall webworm *Hyphantria cunea.* One of the most potent factors is natural selection acting on the number of degree-days (°D) required for development of the insect (Figure 4.3). When the summer was warm, individuals that required high °D were selected for because they pupated late and less of their food reserves were used before cold weather led to reduced metabolic rates. In the following spring more energy remained for conversion to eggs than in individuals requiring fewer °D. A population with high °D requirements was selected for, and selection was reinforced in each successive warm summer. However, if the summer were cool the pupal stage would be reached only by a very small number of individuals before temperatures sank to below the develop-

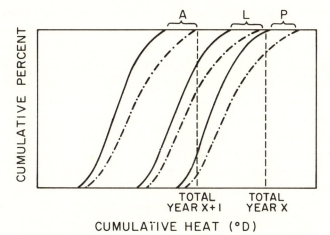

FIGURE 4.3. Hypothetical occurrence of adults (A), fifth instar larvae (L) and pupae (P) of the fall webworm in the field during a warm summer (year X and solid lines) followed by a cold summer (year X + 1 and dash-dot lines), in relation to the cumulative °D above the developmental threshold. In the warm summer all but a minute fraction of the larvae obtained enough heat to pupate: 98 percent of larvae pupated in the cumulative heat total for year X. But in year X + 1, without natural selection, cumulative °D were sufficient for only 20 percent of the larvae to pupate. But in the warm year X, warm-adapted larvae had been selected for, thus retarding the development times (from solid to dash-dot lines). The resulting mortality is double that if natural selection had not occurred.

mental threshold, and high mortality resulted (up to 95 percent). But population decline continued in the following year if a second summer were cool because adults emerged late and many of their progeny were warm-adapted. Thus, the rapidly changing environment imposed strong selection on *Hyphantria cunea*, resulting in populations that were constantly and rapidly evolving. Evidently the cost of evolving was paid in return for the maintenance of high variability in °D requirements. An alternative avoiding this kind of cost would be for individuals to find less

89

variable environments, in the extreme case to become endo-parasitic or to utilize homeothermic hosts. Once the niche is less variable the major selective forces are likely to result from the interaction between host and parasite. Then mutation may provide sufficient, and the best kind of variability for evolving in this milieu, and sexual reproduction can be forsaken.

A numerical example synthesizes several of the ideas expressed in the foregoing paragraphs. The onion thrips *Thrips tabaci* will be used because Lewis (1973) brought together much relevant information on this species. It can be found in some populations with a sex ratio of 1:1 in which amphimixis is presumably prevalent, and in others where males are absent so thelytokous parthenogenesis must be the rule (see Table 4.2 for life history data and population characteristics). Some of the typical attributes of a parasite discussed in previous paragraphs are illustrated: high reproductive potential resulting from short generation time and moderately high fecundity, with the capability for massive increases in populations per host and per unit area. But we are not interested in mean population sizes of

TABLE 4.2. Life history and population characteristics of *Thrips tabaci* (data in Lewis, 1973, from several original sources).

Duration from egg to adult	14 days
Adult longevity	20 days (30°C) 50 days (21°C)
Preoviposition	3 days (21°C)
Mean fecundity	80 eggs (18°C)
Mean fecundity per day	1.8 eggs (18°C) (max 9)
Range in fecundity	12-109
Number of generations per year (in Japan)	at least 10
Mortality per generation before reproduction (from life table on *Limothrips denticornis*)	90%
Peak number per host plant (in agriculture)	3,600 nymphs
Peak population per hectare (in agriculture)	37-74 million
Increase in population during breeding season (i.e. mean for all females)	up to 3,000 times

thelytokous organisms so much as in the capacity of single clones to perpetuate themselves and the adaptive features of one clone versus another.

Suppose a single thelytokous female colonizes a new patch of onions and represents generation one in the year. After 10 generations we may reasonably expect that her progeny are represented by about 1.3×10^8 individuals. This assumes that 10 percent of her progeny reach reproductive age (based on a life table on another species) and lay 80 eggs each. The estimate is conservative for the fittest genets as it uses mean rather than maximum fecundity and mean survival in a population, whereas clones are likely to show great differences in survival.

The amount of new variation present in these progeny can be crudely estimated. If we accept that *Thrips tabaci* is similar to *Drosophila melanogaster* in the amount of genetic information per cell and that Dobzhansky et al. (1977) are correct in assuming that there are 10,000 pairs of genes in *Drosophila* with an average mutation rate per gene per generation of 10^{-5}, then each zygote will contain 0.2 mutations on the average. Thus, the genotype of the original female *Thrips* that leaves 1.3×10^8 progeny after 10 generations will have 2.6×10^7 new mutations in that generation alone on which natural selection can act. In addition there will be what remains of the mutations in the 1.9×10^7 progeny of previous generations that have carried a total of 3.8×10^6 gene mutations. Short generation times such as in *Thrips tabaci* increase the amount of mutation in a season by only about 15 percent compared to that in long-lived females that leave in a single generation a similar number of progeny at the end of a breeding season (e.g. *Ascaris lumbricoides*), but the proportion of favorable mutations contributed by previous generations will be much greater since selection will have weeded out the deleterious mutants. If the 2.6×10^7 mutations are distributed at random among the progeny, 294 individuals would carry as many as 5 to 7 new genes and 1.06×10^8 progeny would have none. This variation will have been generated while the host plant population, say of onion, has hardly evolved because all is produced within a single year, while the generation time of onion is two or three years. Moreover, because of vegetative reproduction through bulbils and sets, one genet may remain in a patch for several years. These clones may be exploited best by individuals and their progeny that carry no new genes. Wherever a new resistance gene arises by mutation in an onion population, a new virulence gene is likely to appear in any coadapted clone of *Thrips tabaci* of superior fitness within a year or two simply because of the large number of progeny produced and the large number of new genes that they contain. *Puccinia graminis* in Australia reproduces asexually since barberry is exceedingly rare, but this pathogen adapted to the new resistant races of wheat as rapidly as they were produced (see Day, 1974 for summary). Indeed, it is possible that virulence genes are present in a population before resistance genes in the host population are present, as was found in Britain for the blight fungus of potato *Phytophthora infestans*, which reproduces asexually there (Shattock, 1977).

A female *Thrips tabaci* from a population with sex ratio of 1:1 may colonize a new patch of onions after being fertilized in the parental patch. After 10 generations she will have been replaced by 6.7×10^7 progeny, assuming similar conditions to those for the parthenogenetic founder, and these may contain 3.4×10^6 new mutations. But in a sexually reproducing population Dobzhansky et al. (1977) estimated that the amount of variation is about 5,000 times greater than that resulting from mutation in each generation. After a few generations in the new patch when an optimal genotype for that patch has appeared, the recombinational load will be enormous and may cause extinction in that patch. Of course, for an inbreeding haplodiploid species like *Thrips tabaci* we must modify the estimate of variation in a population. Haplodiploid species have about half as many heterozygous loci per individual as other insects and half as many polymorphic loci per population (Selander, 1976). The founding female may well be homozygous (and carry sperm of a similar homozygous genotype) and recombinational load in subsequent generations will be slight; variation will be comparable to that in the thelytokous female. But arrival of another individual in the patch, unrelated to the founder and her mate may introduce enough new variation to produce a dangerous amount of recombinational load.

Maynard Smith (1978) and others have argued cogently against the view that parthenogenesis is a viable long-term strategy or that it is in any way creative in the sense of leading to the evolution of new taxa after the parthenogenetic clone has arisen. The lack of any large thelytokous taxa, other than the bdelloid rotifers, clearly supports this view, but this fact is not sufficient to discount the creative possibilities in parthenogenetic forms on a smaller scale or their longevity as morphospecies. The commonness of parthenogenesis is usually underestimated because detailed studies are lacking. Thirty-two percent of plant-parasitic and soil-dwelling nematodes with known modes of reproduction listed in Table 4.1 are reported to be obligately parthenogenetic. In North America 17 of 33 species of Lumbricidae reproduce principally by parthenogenesis (Jaenike and Selander, 1979). Maynard Smith (1978) states that the insect order Hymenoptera has only a few thelytokous species, but in the British fauna in one hymenopterous family, the Tenthredinidae, 13 percent of the species are known to be thelyokous at least in some parts of their range (see the 48 species listed in Benson, 1950 and the total number of 358 species in the British Isles, Table 2.1). Moreover, one genus *Nematus* contains three parthenogenetic species with no known closely related, sexual relative (Benson, 1950). Either these morphospecies have arisen from a parthenogenetic stock, or they have outlived a sexual progenitor, both possibilities contravening the conventional wisdom. Some relatively large genera exist also in which parthenogenesis is the major mode of reproduction as in *Meloidogyne* and *Pratylenchus* (Nematoda) (Franklin, 1971; Triantophyllou, 1971), genera in the Superfamily Mermithoidea (Triantaphyllou, 1971), which are parasitic nematodes of invertebrates, and the nonparasitic roach genus *Pycnoscelus* (Roth, 1967, 1974; Roth and Cohen, 1968). In the last case the parthenogenetic clones with different karyotypes can be regarded as typological species. Within a karyotype several distinct clonal types may exist in a region, and, although Parker et al. (1977) suggest multiple invasions from the parental stock, the alternative explanation that new typological species have arisen from the parthenogenetic stock cannot be discounted.

Several authors have recognized the superiority of sex for surviving the biotic forces of predation, parasitism, and competition (Van Valen, 1973; Levin, 1975; Jaenike, 1978; Maynard Smith, 1978; Glesener and Tilman, 1978), and competition has been shown to select for high levels of genetic variation (Powell and Wistrand, 1978). However, parasites commonly live in non-equilibrium conditions (Chapter 3) with superabundant resources where competition is relatively unimportant (Chapters 5 and 6), and for internal parasites at least, predation is unlikely to be a major selective factor in most cases. In this chapter I argue that genetic heterogeneity of hosts is generated at rates sufficiently low to permit rapid tracking by parasites even with a parthenogenetic mode of reproduction. Thus the arguments for selection of sexuality in biotically diverse environments do not carry as much weight for parasitic species as for free-living species.

White (1973:745) certainly recognized the adaptive nature of thelytokous parthenogenesis but concluded that it was rare because of "the scarcity of sufficiently narrow, invariable ecological niches." I argue in this monograph that such niches are superabundant for parasitic organisms, that parthenogenesis is common in parasites, and that the subject deserves much more study before current conventional wisdom congeals into dogma.

The population structure of parasite species envisaged in this chapter and in the ecological concepts given in Chapter 2 is that of a species split into many small populations with little gene flow between populations. (Small refers to the mean population size of many patches and many generations, not to the epidemic condition we usually perceive.) In obligate thelytokous species isolation of each clone is absolute. Founding events followed by population eruptions are common and inevitably followed by population crashes of equal frequency and extinction in many patches. The founder effect (see Mayr, 1963) will operate in space and time. Colonization of a new patch will be achieved in many cases by a single individual. The flush-crash cycle within a patch may leave a single survivor or very few to perpetuate the population or clone (Carson, 1968, 1975). While numbers are small, genetic drift will probably play a major evolutionary role in amphimictic populations. Note that the probability of random fixation (p) of a gene or chromosomal mutation decreases as population size increases as follows:

$$p = \exp\left(-s[N-2]\right)/2N$$

where N is effective population size and s is the heterozygote disadvantage (Nei, 1975). But a compensating factor in populations with skewed sex ratios in favor of females as in many populations of parasitic species (e.g. arrhenotokous species and nematodes) is that the effective population size (N) decreases as the skew increases:

$$N = 4FM \;/\; F + M$$

where F is the number of females and M the number of males (Wright, 1940). Thus if 1 male mates with 100 females, N is approximately 4.

Differences in host populations or host species between patches and the environments in which they exist will impose intensive differential selection in each situation. Wright (1931) and Wade (1977) recognized the evolutionary potential through group selection of species in which local extinctions occur relatively frequently, followed by recolonization. In elegant laboratory experiments Wade (1977) found that under the influence of group selection, evolutionary changes were rapid and large, even occurring in two or three generations, and could occur in the presence of considerable gene flow between populations. Under these influences differentiation between populations will be considerable relative to the confines of the prevalent breeding system and mode of reproduction in these populations. Even in relatively uniform environments, such as glass houses, population differentiation proceeded to such an extent that genetic incompatibilities between individuals from different glass houses were very common (Helle and Pieterse, 1965). During the founding phase and early colonization when populations are very small chromosomal rearrangements become readily fixed and result in isolation between populations (see Wilson et al., 1975; Bush et al., 1977 for consideration of such influences on vertebrates). The glass house mite studied by Helle and Pieterse (1965) appeared to be isolated by chromosomal differences between populations.

Differences between available patches for a parasitic species may have profound effects upon many of the attributes of a species. Such effects were apparent in the milkweed longhorn beetle *Tetraopes tetrophthalmus*, whose larvae feed most commonly on and in the roots of the common milkweed *Asclepias syriaca* (see Price and Willson, 1976 for details). But one population of about 50 adult beetles was found on the smaller horsetail milkweed *Asclepias verticillata*. A comparison between this population and one about 250 meters away on the major host revealed that they differed significantly in every character studied: phenology, beetle size, longevity, ovariole number, egg size, and mating preferences. Ecological isolation between populations was evident because each host species occurred over a different range on a moisture gradient and vegetation around the smaller host provided a mechanical barrier to colonization by flight. The stage was set for rapid differentiation of the newly established population for existence on a drier site and a different diet. Many such colonization events may go unnoticed because of their very local nature and the inconspicuous numbers involved, but they are of profound importance in the evolutionary biology of parasites. The beetle population on *A. verticillata* may be on the brink of extinction, but if it persists, as somewhere one will, it may evolve into a new small species of *Tetraopes* on a new host, just as *Tetraopes linsleyi* seems to have specialized on *A. subverticillata* in western North America (see Hovore and Giesbert, 1976).

The demands on a population of parasites imposed by

97

colonization of a new host species and the consequently new ecological conditions may well be sufficient to select for specialization on a particular host, with many populations exploiting different hosts. The parasite species may appear to be polyphagous but in fact many populations are monophagous. Several examples of this have been found among the butterflies (Gilbert and Singer, 1975; Singer, 1971). Once different hosts are utilized, sufficient ecological isolation may be present for speciation to result. A progenitor *Schistosoma* species seems to have speciated into *S. haematobium* on savannah primates and *S. intercalatum* on forest-dwelling primates (Wright and Southgate, 1976). When the microfilarial parasite *Loa loa* colonized man from arboreal primates, the available vectors automatically changed from crepuscular and nocturnal, canopy-dwelling tabanid fly species to other species that fly in open sites during the day (Duke, 1972). The host shift immediately caused very effective ecological isolation between populations.

Host shifts may be readily achieved, also in sympatry, by colonists that are reproductively isolated from members of the parental population. Bush's model for sympatric speciation, initially proposed in 1969 and improved upon in 1974 and 1975 (1975a), is based on his studies of phytophagous parasitic insects in the family Tephritidae, but he recognized the possibility that the model may apply to many obligate parasites on both plants and animals (Bush, 1974, 1975a). The model provides the genetic mechanisms for a host shift from apple to cherry by the apple maggot *Rhagoletis pomonella*, which occurred in about 1966 in Wisconsin. The host shift is accomplished by individuals carrying only two mutations, one for host selection and one for survival in the host. As suggested earlier in this chapter, mutation probably provides the necessary material for evolutionary events while recombination is of less importance. However, recombination would increase the probability of the two mutant genes occurring in one individual.

Bush envisions a race of the fruit fly on apple with a host selection gene H_1 and host survival gene S_1, both in a homozygous state. Somewhere in the population a mutant gene H_2 is produced that changes host selection from apple to cherry. Another mutation at the S locus to S_2 preadapts the carrier to survival in cherry. When the two mutant genes occur together in homozygous condition, the carrier selects cherry as a host and its larvae are able to survive in cherry (Figure 4.4). A new host will be colonized and reproduc-

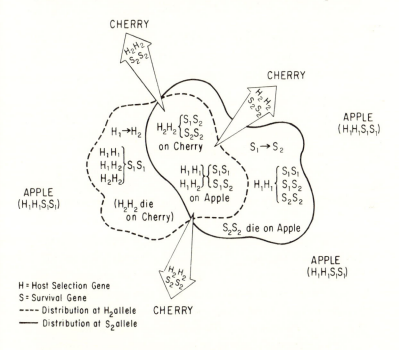

FIGURE 4.4. Hypothetical genetic model for the sympatric speciation of *Rhagoletis pomonella* into the parental species remaining on apple and a new species on cherry (from Bush, 1975a).

tive isolation is likely to exist between the new race and the parental race for several reasons.

(1) There is evidence that larval conditioning influences host plant selection by adults (Huettel and Bush, 1972).

(2) Cherry fruits are available for oviposition much earlier than apples, so there will be strong selection for the phenology of the cherry race to be advanced. Allochronic isolation will become more effective.

(3) Heterozygous H_1H_2 individuals may oviposit on either host but are likely to select the host from which they emerge because of conditioning and phenology.

(4) Some gene combinations would be lethal such as $H_2H_2S_1S_1$ on cherry and $H_1H_1S_2S_2$ on apple. Any heterozygous individuals would have lower fitness than the homozygotes since some lethal combinations would appear in the next generation.

Reproductive isolation would increase with time, and, as Bush (1975a) says, the decision on when the speciation event has occurred is arbitrary. Certainly there should be no doubt that speciation has occurred when hybrids on the two races have lower survival on each host than does the appropriate parent, as in the hybrids of *Puccinia graminis tritici* and *P. graminis secalis* studied by Green (1971). In this case divergence of gene pools has occurred even though both species utilize barberry as the spring host. Parasite populations must adapt rapidly in many ways to the host, and every adaptive route will increase isolation. For example, *Xanthomonas malvacearum*, which causes cotton bacterial blight, had more antigens in common with its host plant *Gossypium hirsutum* than with a congeneric bacterium (Schnathorst and DeVay, 1963), and, as Deverall (1977) points out, serological relatedness is normally thought of as an indicator of taxonomic affinity. There is a known gene-for-gene relationship between this bacterium and its host (Brinkerhoff, 1970). When *Loa loa* colonized man and so changed from transmission by a crepuscular and nocturnal vector to a diurnal vector, there was a strong selection pressure, in order to be included in the vector's blood meal, for microcilariae to migrate into the peripheral blood of humans during the day instead of in the evening that reinforced the ecological isolation between hosts and vectors (Duke, 1972).

Bush has provided a plausible genetic mechanism for sympatric speciation, and the view long held by many biologists that such speciation is more likely in some taxa than allopatric speciation thereby receives strong support. For example, students of the Homoptera like Evans (e.g. 1962) and Ross (e.g. 1962) felt that closely related and sympatric species were too numerous for them to be accounted for by the allopatric model. Anyone finding 43 species in a single beetle genus on a small island in the Pacific Ocean, as Zimmerman (1938) did, would suspect sympatric speciation as the cause. Ross (1962) anticipated Bush's concern for the characters that predispose a species for sympatric speciation. Bush (1975a) lists the following preadaptive traits for phytophagous parasitic insects: (1) Mating occurs on or near the host plant. (2) The adult female selects the host while the larvae have no choice. (3) The species are monophagous or stenophagous, attacking groups of closely related host plant species. (4) Host selection is under genetic control. (5) Larval survival is dependent upon the action of survival genes. (6) Species are univoltine with genetic control of emergence time. (7) Host plant induction (conditioning) is under genetic control. Several of these traits are evident in the codling moth *Laspeyresia pomonella* that appears to be speciating sympatrically (Phillips and Barnes, 1975).

As White (1973:336) says, "It does not seem very likely

101

that the same model of the speciation process should apply equally to oceanic birds, pelagic fishes, small rodents, land molluscs, *Drosophila*, monophagous gall-forming insects, hermaphroditic and parasitic flatworms etc." Population structure will differ considerably from one kind of species to another. White emphasizes that even closely related species can differ karyotypically and that chromosomal races frequently may be found in species with relatively sedentary individuals. With frequent founding events individuals carrying chromosomal rearrangements may establish new populations reproductively isolated from the parental population. White states that in most cases where chromosomal rearrangements have been influential in speciation the karyotype changed at an early stage in evolutionary divergence. Thus given the population structure envisioned here for many parasitic species, we should expect speciation through chromosomal rearrangement to be common. This appears to be true for speciation of the small parasitic chalcidoid wasps (see Goodpasture and Grissell, 1975) in which karyotypes differ significantly between closely related species (Goodpasture, 1975). Blackman (1977) found two sibling species of aphid in the genus *Euceraphis*, one on each of the related hosts *Betula pubescens* and *B. pendula*. The difference in karyotype with one probably derived from the other by a simple fusion suggests that the two-gene model of Bush does not apply. The host races of *Rhagoletis pomonella* and other members of the species group all have the same karyotype (Berlocher, 1976). Triantaphyllou (1971) recognized the importance of karyotype change in nematode speciation. Establishment of polyploidy and parthenogenesis is particularly common in the plant parasitic nematodes. Craddock (1974) identified 10 chromosomal races of the exceptionally sedentary wingless stick insect *Didymuria violescens* in southeastern Australia and concluded that parapatric speciation was the most likely mode. She suggested that chromosomal changes as isolating barriers would actually stimulate the evolution of reinforcing reproductive isolating mechanisms and thereby accelerate the divergence of populations, compared to the passive divergence of geographically isolated populations. This is because selection against interracial hybrids would tend to favor the evolution of premating isolating mechanisms.

Craddock (1974) and White (1973) comment on the probable rarity of chromosomal rearrangements that improve the fitness of the carrier over the parental karyotype either in the same environment or in an adjacent one. But it must be emphasized that through many years such a vast number of individuals in a parasite species or clone will be produced that even with exceedingly low probabilities for such karyotypic change their eventual occurrence is inevitable. And as White (1973:761) comments, "Every chromosomal rearrangement, every evolutionary change in a karyotype, is a genetic revolution; and once it has occurred, a whole series of point mutations will be tried out in novel combinations in the new karyotype."

In order to summarize this chapter a definition will be attempted of the kind of species a parasite is likely to be. The definition will obviously not apply to all parasites but I feel that it will represent a modal position around which great variation may be observed. Also, to be realistic a great range in some criteria must be acknowledged. The summary (Table 4.3) follows Mayr's (1963) Table 14-1 titled "Classifying criteria for kinds of species." He claims no attempt at a complete list, recognizing that almost all species characteristics are significant and that inevitably each category overlaps to some extent with others. Each criterion is

TABLE 4.3. A tentative definition of the kind of species parasites are likely to be compared to free-living, relatively large organisms (e.g. some predators).

Criterion	Kind of species	
	Parasites	Predators
1. System of reproduction	Biparental sexual reproduction to mitotic parthenogenesis with a strong tendency toward hermaphroditism and parthenogenesis	Biparental sexual reproduction
2. Amount of gene flow	Largely inbreeding	Outbreeding
3. Size of populations (demes)	Usually small to very small but erupting to very high numbers at times	Large and fluctuating to a lesser extent
4. Phenotypic plasticity	Polymorphic ranging to monomorphism even between species (siblings)	Monomorphic, without sibling species
5. Sequence of generations	Usually rapid	Slow
6. Environmental tolerance	Narrow	Broad
7. Difference in origin	Rapid by sympatric (gene mutations) or parapatric (chromosomal mutations) speciation. Instantaneous by polyploidy	Slow by allopatric speciation
8. Rate of evolution	Rapid	Slow
9. Structure of species	Often polytypic	Monotypic
10. Variation in chromosome number	None in many species to much in morphospecies such as *Didymuria violescens*.	Little variation or none
11. Degree of intra- and interspecific fertility.	Great range in intra- and interspecific fertility	High intra-specific, low interspecific fertility
12. Presence or absence of hybridization in nature	Interspecific hybrids found rarely	Usually none
13. Pattern of distribution	Very local to widespread	Widespread

stated in relation to free-living, relatively large organisms such as some predators, although some criteria will be seen not to be effective in distinguishing the two kinds of organism.

CHAPTER FIVE

Adaptive Radiation and
Specificity

What is the potential for further adaptive radiation in parasite taxa? Can we predict to what extent speciation will continue in different groups of parasites? How does the specificity of parasites modify the amplitude of adaptive radiation?

These questions can be approached by asking another simple question: have parasite taxa evolved to utilize all available resources provided by the host group to which the parasites are adapted? If we accept the Hutchinsonian definition of the niche (Hutchinson, 1957), we can ask if vacant niches exist. Extrapolating from Ecological Concept 3 that predicts unsaturated communities in ecological time, we should also anticipate that communities remain unsaturated in evolutionary time, that species packing is very loose, and that many resources remain unexploited. This statement will be tested in the following paragraphs and in Chapter 6.

Evolutionary Concept 2(i) stating that the extent of adaptive radiation is related to the diversity of hosts available is intuitively obvious and well supported in the literature. Six examples of the trend were provided in Chapter 2. An important feature common to these relationships is that the slope of the regression is much less than one (Table 2.3), except for Strong's (1977a) data on hispine chrysomelids that use only common hosts. A relatively small proportion of the apparently available hosts are utilized by parasites of a certain taxon. When parasites specialize in body regions as do many ectoparasitic insects, mites, and agromyzid flies and so create the possibility for stable coexistence of many more parasitic species than host species, the low number of parasites compared to their hosts becomes even more evident. Why is this?

The answer must lie in one of the three following possibilities.

(1) Parasite communities may be in equilibrium as defined in island biogeographic theory (MacArthur and Wilson, 1967) but the equilibrium number of species in evolutionary time is very low because of low colonization rates and high extinction rates.

The other two possibilities invoke non-equilibrium conditions.

(2) Many host islands are too small (in area occupied or population size) or too distant (geographically, ecologically, biochemically, taxonomically) to be colonized extensively in evolutionary time, so hosts are not fully exploited and communities remain unsaturated.

(3) Many communities remain unsaturated in evolutionary time because although the hosts are common and proximal to sources of colonists, they have remained uncolonized: there is much room for further adaptive radiation.

We must determine which hypothesis is supported by the available data.

The answer to this question may also help to resolve the apparent inconsistency between Ecological Concept 3 emphasizing non-equilibrium conditions in parasitic communities and the evidence cited in Evolutionary Concept 2(ii) showing good species-area relationships between parasites and hosts that support the concept that parasitic communities are in equilibrium. The answer will obviously contribute to the debate on whether time or host area is critical in determining the number of species per host; whether parasite species accumulate nonasymptotically or whether the host area imposes an asymptote to species richness (e.g. see Strong, McCoy, and Rey, 1977). Strong's important contribution to this debate that has used a broad approach should be complemented by more detailed studies restricted to a guild of parasites. The benefits of the narrower view are at least threefold.

(1) The niche relationships between species and species packing can be understood, so the number of discrete resources available on a host species can be estimated. Thus the maximum possible number of parasite species can be estimated. Therefore the maximum number of parasite species per host can be anticipated for hosts that are abundant and predictions made about numbers of parasites in saturated communities.

(2) Species-area relationships between parasites and hosts can be studied on a comparative basis, for example, to compare species richness of the parasite guild on one host group with that on another in the same region.

(3) All species in the guild, both common and rare, can be included in the analysis, whereas commonly only pest species, that is very abundant species, have been recorded and used. We know little about the variance generated by the inclusion of innocuous species.

Strong (1977a, b) has already taken this more detailed approach in his studies of hispine beetles on *Heliconia latispatha* and other Zingiberales hosts.

The leaf-mining flies in the family Agromyzidae are sufficiently well studied in Britain so that the data available satisfy the above criteria and provide tentative decisions on the three hypotheses accounting for species richness of parasites on their hosts (see Lawton and Price, 1979 for details). The host relations of agromyzids in Britain are listed in Spencer (1972) together with the resources utilized by each species: leaf blade miner, leaf midrib miner, stem miner, stem borer (including gall makers in stems), seed feeder, and root feeder, being the types of feeding. Spencer (1972) lists all species of agromyzid in Britain, both common and rare. The analysis is restricted to agromyzids on the family Umbelliferae so the species-area relationship between these taxa can be compared with agromyzids on other plant families recorded by Spencer. Host plant areas are provided in Perring and Walters (1962) as the number of ten-kilometer squares in Britain in which the plant occurs.

The relationship between host species area and number of agromyzid species per host is significant, but accounts for only 32 percent of the variance (Figure 5.1). This contrasts markedly with the high r^2 values obtained by Opler (1974), Dritschilo et al. (1975), and other studies listed in the upper

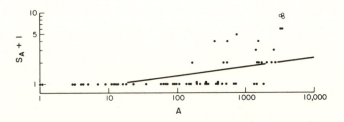

FIGURE 5.1. Regression of the number of agromyzid species, $S_A + 1$, on the area occupied (A) (number of ten-kilometer squares) by host plants in the family Umbelliferae in Britain. For agromyzids on umbrellifers (●), the regression equation is $\log_n (S_A + 1) = 0.13 \log_n A - 0.41$; $F_{1,59} = 27.70$; $p < 0.001$. Agromyzids on *Senecio*, *Taraxacum* and *Ranunculus* (o) are not included in regression (after Lawton and Price, 1979)

part of Table 5.1. But the result falls within the range found in other studies by Strong (1974c), Cornell and Washburn (1979) on Atlantic oaks, Strong and Levin (1975), and others (Table 5.1). Clearly when the species-area relationship accounts for only 50 percent or less of the variance, other major influences on species richness must be sought. The results so far show a wide spectrum in the influence of area on species richness, so we cannot easily accept the claim that the asymptotic model, with the asymptote defined by host geographic range, holds in all cases (see Strong, McCoy, and Rey, 1977).

TABLE 5.1. Host-parasite systems in which species-area regressions have been calculated and an estimate of the percentage variance accounted for (r^2 x 100) has been obtained. In each case the independent variable was area occupied by the host species and the dependent variable was the number of parasite species per host species. Note that studies account for 90% to less than 1% of the variance. Asterisks indicate studies on area differences within a host species; all others compare area differences among host species.

Parasite	Host	(r^2 x 100)	Source
Leaf miners	Oaks	90	Opler (1974)
Mites	*Peromyscus* spp.	86	Dritschilo et al. (1975)
Insects	Woody shrubs	85	Lawton & Schröder (1977)
Arthropods	*Astragalus sericoleucus	71-85	Tepedino & Stanton (1976)
Cynipids	Oaks (California)	72	Cornell & Washburn (1979)
Insects	Perennial herbs	71	Lawton & Schröder (1977)
Insects	Trees	61	Strong (1974a,b)
Insects	*Sugarcane	61	Strong, McCoy, & Rey (1977)
Insects	Annual plants	59	Lawton & Schröder (1977)
Insects	Monocotyledons	51	Lawton & Schröder (1977)
Insects	*Cacao	47	Strong (1974c)
Mites	*Microtus* spp.	46	Dritschilo et al. (1975)
Cynipids	Oaks (Atlantic)	41	Cornell & Washburn (1979)
Mites	Cricetid rodents	37	Dritschilo et al. (1975)
Parasitoids	Umbelliferae	35	Lawton & Price (1979)
Agromyzids	Umbelliferae	32	Lawton & Price (1979)
Fungi	Trees	28	Strong & Levin (1975)
Microlepidoptera	Umbelliferae	24	Lawton & Price (1979)
Arthropods	*Phlox bryoides	10-16	Tepedino & Stanton (1976)
Leafhoppers	Trees	.04	Claridge & Wilson (1976, 1978)

The following are notable features in Figure 5.1.

(1) Many plant species have no agromyzid parasites although they occupy an area as large as others that have parasites.

(2) Several common plant species support only one parasitic species whereas species of similar range support four and five times as many.

(3) The two species of plant with the widest range, *Heracleum sphondylium* and *Angelica sylvestris* each with five agromyzid species, have fewer parasites than plants outside the Umbelliferae with similar ranges: *Senecio jacobaea*, *Taraxacum officinale*, and *Ranunculus acris*, each with eight agromyzid species.

(4) Plant hosts recorded from less than 160 ten-kilometer squares support no agromyzid species.

(5) High numbers of agromyzids on some hosts cannot be fully accounted for by their association with closely related species making their effective geographic range greater. A reanalysis taking into account such taxonomic and biochemical affinities based on information in Heywood (1971) yields only a slightly better species-area relationship ($r^2 = 0.34$).

The relationship depicted in Figure 5.1 is not sufficient to discount the existence of equilibrium conditions in agromyzid communities. Competitors, such as microlepidoptera and other leaf-mining taxa, may preempt resources and complement the species richness of agromyzids. This does not appear to be the case. The microlepidoptera on British Umbelliferae show a species-area relationship very similar to the agromyzids ($r^2 = 0.24$) (Figure 5.2). A regression of agromyzids plus microlepidoptera on host area accounts for

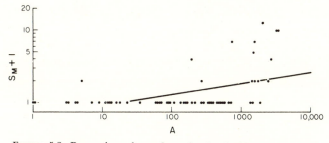

FIGURE 5.2. Regression of number of microlepidoptera species, $S_m + 1$, on the area occupied (A) (number of ten-kilometer squares) by host plants in the family Umbelliferae in Britain. Regression equation is $\log_n (S_m + 1) = 0.16 \log_n A - 0.49$; $F_{1,57} = 18.80$; $p < 0.001$. (after Lawton and Price, 1979).

very little of the variance $(r^2 = 0.34)$ leaving most of the variance unexplained by host plant area. In fact a positive relationship exists between number of species of agromyzids on a host and number of microlepidoptera species in Britain $(r^2 = 0.47)$, and a similar relationship is seen between agromyzids and all other leaf miners in Europe $(r^2 = 0.32)$. It appears that each group of parasites is colonizing host plants in a similar way and independently of each other, and this in turn suggests that many resources are available for colonization and these have yet to be fully utilized.

The equilibrium model may still hold if host species are unequally resistant to colonization because some are more favorable than others to parasites and predators of agromyzids. Plants may live in sites that are more favorable to agromyzid enemies than others (e.g. high humidity for parasitoids), and many Umbelliferae are known to be favored sources of nectar for parasitoids. A potential host may be associated with another host that harbors several agromyzids and their enemies, and the overflow of species may maintain very high extinction rates on the associated species. The parasitoid records on Agromyzidae provided by

111

Griffiths (1964-1968), Fischer (1962), Graham (1969), and Bouček and Askew (1968) enable a partial test of this possibility. Most of these records provide not only the host of the parasitoid, but the plant host of the agromyzid, so parasitoids can be associated with plants and a species-area relationship generated for parasitoids and plants (Figure 5.3).

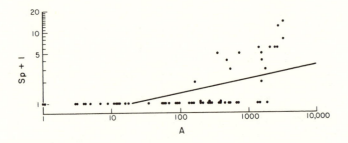

FIGURE 5.3. Regression of number of parasitic braconids, S_p + 1, on the area occupied (A) (number of ten-kilometer squares) by members of the Umbelliferae in Britain with which they are associated. Regression equation is $\log_n (S_p + 1) = 0.20 \log_n A - 0.60$; $F_{1,59} = 31.82$; $p < 0.001$ (after Lawton and Price, 1979).

The parasitoids show a similar relationship to the agromyzids and provide no evidence of complementarity ($r^2 = 0.35$). There is a positive correlation between the number of agromyzids on a host plant species in Britain and the number of parasitoids associated with the host plant ($r^2 = 0.35$) and also between the number of agromyzids per host genus in Europe and the number of parasitoids with that genus ($r^2 = 0.34$). The parasitoids have responded to the Umbelliferae in a way very similar to the herbivores and appear not to be responsible for blocking colonization in a major way.

Lawton and Price (1979) have considered several other factors that influence colonization by agromyzids in evolutionary time. Plant size, leaf form, and whether plants were terrestrial or aquatic were the only characters that improved the relationship between plant area and number of parasite species. Even then about 50 percent of the variance was left unaccounted for.

Based on the evidence available at present, the conclusion must be that agromyzid communities are not in equilibrium. Both non-equilibrium hypotheses are valid: many potential hosts with small geographic ranges remain uncolonized, but also many common plant species have resources unutilized by agromyzid species. Given enough evolutionary time, there is no apparent reason why the latter group should not continue to accumulate species up to an asymptotic, equilibrium level.

Equilibrium levels may be much higher than the number of agromyzid species found on any umbellifer in Britain today. Among the leaf blade miners, by far the largest group of agromyzids, three species are known to coexist on the same host (e.g. *Heracleum sphondylium, Pimpinella saxifraga*), and six and seven species on one host in other plant families is common. Four leaf midrib miners occur on *Taraxacum officinale* and three stem borers on *Senecio jacobaea*. Two stem miners are found on *Melandrium rubrum* and three on *Galium mollugo*. Ignoring the other resources rarely utilized by agromyzids, we could conservatively expect 12 to 15 species of agromyzid on hosts with the largest geographical range, once equilibrium conditions are finally reached. Competition for plant resources, as envisioned by Janzen (1973a), is unlikely to reduce this number since patchiness of resources, low colonization rates, and the usually low level of resource utilization on any one plant, will dilute the intensity of interactions even when parasite species occur in the same vicinity.

The example of the Agromyzidae is not an isolated one.

113

Many other parasite taxa underexploit the available host resources as seen in the microlepidoptera in Britain and the other examples where the regression of parasite numbers on potential host numbers yields a slope of much less than one. The possibility exists that not all hosts are available to herbivores. For example, no agromyzids are found on the 25 species of Ericaceae in Britain, 4 of which are very common. Nowakowski (1962) noted that plants with xeromorphic leaves, as in the Ericaceae, are avoided by agromyzids. Much more research is needed on accessible and nonaccessible hosts for colonization by parasites, but until this is understood for any taxon, the parsimonious view must be accepted that most hosts are available and many are underexploited by parasites.

Why, then, should there exist significant species-area relationships between hosts and all parasites within a large taxonomic unit such as the Class Insecta or the Division Mycophyta (fungi)? In certain cases there is good evidence for equilibrium conditions, particularly in the leaf-mining lepidoptera on California oaks (Opler, 1974), but this is a narrowly defined taxonomic unit. In other examples innocuous or rare species have not been included in the regression that would probably add considerably to the variance of the relationship. Pest species are frequently generalists and will therefore colonize in ecological time, whereas specialists need evolutionary time in which to colonize. The distinction should be made in analyses. Where significant species-area regressions have been obtained, there are usually enormous differences between numbers of parasites on hosts with geographical ranges similar in size, indicating unutilized and undersaturated resources in many cases. These differences need to be understood. The answers are not readily available, but the need for more detailed analysis of large parasitic assemblages is apparent.

A rather different approach to studying available resources and parasite utilization using parasitic wasps in the family Ichneumonidae yields the same conclusion as with the Agromyzidae. There is a vast matrix of unexploited host resources available for future adaptive radiation.

A strong correlation exists between fecundity of ichneumonid parasites and the probability of survival of their hosts, as estimated from the survivorship curve (Price, 1974a, 1975b). Therefore the accumulation of many host survivorship curves can generate a matrix of resources, and the extent to which these are exploited, indicated by the range of fecundities in the parasite family, can be examined. Most hosts are Lepidoptera, for which the range in shapes of survivorship curves can be obtained from the lierature; a representative group of 20 is illustrated in Figure 5.4. The relative probabilities of survival (*PS*) can be calculated for parasites attacking any one host at any stage because, for most ichneumonids, emergence is from the pupal stage. By assuming that $PS=1$ if the pupal stage is attacked, *PS* at other stages attacked is calculated by dividing the percent of hosts at the pupal stage by the percent at the stage attacked. It is also known that female ichneumonids that attack the pupal stage lay a mean of 40 eggs in a lifetime (Price, 1975b). Therefore, the fecundity required by a parasite attacking any other stage in order to equalize the number of progeny (*P*) it leaves in the pupal stage of the host can be calculated $(P = 40/PS)$. A linear relationship exists in the Ichneumonidae between fecundity and ovariole number (which is more commonly and reliably recorded in the literature), so the conversion from one to the other can be made (Price, 1975b).

The predicted minimum number of ovarioles required for parasites attacking each stage of the host fits closely to those observed in nature (Figure 5.5). However, the maxi-

115

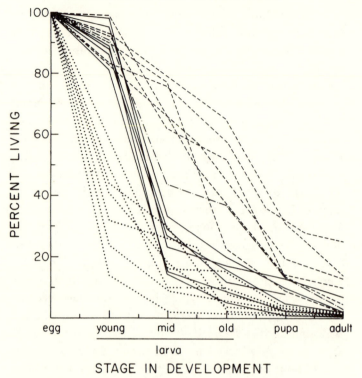

FIGURE 5.4. Surviorship curves of 20 species of Lepidoptera published in the literature, plotted as the percent of the initial cohort alive at each stage of development.

mum number of ovarioles required by parasites attacking any stage but the pupa has not been reached. In fact a very small proportion of the range in ovariole numbers needed to exploit host eggs or young larvae fully has been achieved by the ichneumonids. For these stages to be fully exploited we should see species ranging in fecundity from 128 eggs to 5,245 eggs per female, or 8 to 275 ovarioles per ovary for those attacking young larvae, and from 140 eggs to 10,850 eggs, or 9 to 548 ovarioles per ovary for females attacking

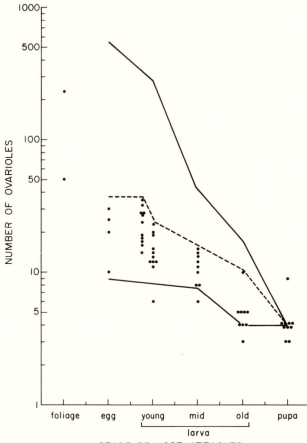

FIGURE 5.5. Number of ovarioles per ovary in 58 species of ichneumonids in relation to the stage of host attacked (data from Price, 1975b; Dowell, 1977; and Quednau and Guevrement, 1975). Solid lines represent predicted maximum and minimum ovariole numbers required to exploit fully all host stages with survivorship as in Figure 5.4. The dashed line represents the upper limit in ovariole number reached so far in the adaptive radiation of the Ichneumonidae.

117

eggs. The highest ovariole number for an ichneumonid parasite known to attack eggs is 30, and for early larvae it is 35 (Price, 1975b). Apparently only 13 percent of available resources have been exploited by parasites attacking young larvae and only 5 percent by those attacking eggs.

If a parasite evolved to lay 10,850 eggs, it would probably be the most fecund nonsocial insect known, but there probably exists a physiological limit to fecundity that will be reached before all the host stages can be fully exploited by the Ichneumonidae. However, the ancestral ichneumonid was undoubtedly of low fecundity, attacking late host stages. Adaptive radiation has, in general, progressed up the survivorship curves of hosts (Price, 1974b), and the physiological limit has not yet been reached. Species exist with 60 and even 230 ovarioles per ovary (Iwata, 1960). The potential resources for further adaptive radiation appear to be even greater than those utilized so far.

SPECIFICITY

The first question posed at the beginning of this chapter has been answered. That further adaptive radiation will be extensive should be disputed by few. But the second question of how extensive this radiation will be is open to considerably more doubt. The two working principles most readily available to answer this question, derived from the general concepts, come from Evolutionary Concept 2 which states that adaptive radiation is extensive and depends upon the diversity of hosts (2.i) and the degree of specialization on hosts (2.iv). Given a certain number of hosts, many more specialists than generalists can pack onto the resources. Radiation should be more extensive among specialists (see also Janzen, 1973b). The other parts of Evolutionary Concept 2 are of less predictive value because we have little idea of what equilibrium numbers of parasites should be for a given host geographic range (2.ii) and evolutionary time available (2.iii) is seldom adequately known.

The ecological forces selecting for specialization or the behavioral and biochemical characteristics of parasites able to utilize a broad spectrum of hosts are little understood and often complex (e.g. see Wood and Graniti, 1976). But, in general, the more closely knit the life histories of parasite and host are, the more specific will be the parasite; this trend is illustrated by parasitic plants in the families Scrophulariaceae and Orobanchaceae (Govier, 1966).

A comparison of the insect parasites of birds and mammals also illustrates the trend (Table 5.2), and host lists for these organisms provide some data for preliminary analysis. (See also Marshall, 1976 for a similar approach.) The chewing lice in the family Philopteridae (Order Mallophaga) are wingless as adults and are ectoparasitic on birds throughout their life cycle. The Mallophaga are known to be very specific in their host relationships (Hopkins and Clay, 1952). Analysis of data provided by Theodor and Costa (1967) shows that 87 percent of the philopterid species in Israel are known from only one bird species. This high specificity must be related to lack of a highly mobile dispersal stage and the entire life cycle being spent on a host. With such extreme specialization, many parasite species should be expected if many host species are available. Therefore, it is significant that the Philopteridae ranked as the sixth largest family of carnivorous parasites in the British insect fauna (Table 2.1), the only family of parasites comparable in size to those parasitic on insects that have many more hosts available.

The Streblidae (Order Diptera) are blood-sucking bat ectoparasites, females of which deposit their fully developed

119

TABLE 5.2. Percentage of species in families of parasitic insects on mammals and birds in each class of number of hosts attacked. From analysis of host-parasite lists for Mallophaga (Theodor and Costa, 1967), Streblidae (Wenzel, 1976; Wenzel, Tipton, and Kiewlicz, 1966), Oestridae (Zumpt, 1965), Hystrichopsyllidae (Hopkins and Rothschild, 1962, 1966) and Hippoboscidae (Bequaert, 1956).

Number of hosts	Philopteridae	Streblidae	Oestridae	Hystrichopsyllidae	Hippoboscidae
1	87	56	49	37	17
2	9	22	19	20	9
3	2	13	7	9	15
4	1	5	6	5	13
5	1	2	6	5	7
6		1	4	6	4
7		1	4	2	4
8			0	3	4
9			4	3	2
10-19			2	9	4
20-29				<1	7
30-39				<1	2
40-49					0
50-59					2
60-69					0
70-79					0
80-89					7
NUMBER OF PARASITES	122	135	53	172	46

larvae on surfaces in the bat roost (Askew, 1971). The pupa and emergent adult thus remain in close proximity to the bat host population and considerably isolated from bats utilizing other roosts. Host-parasite lists in Wenzel (1976) and Wenzel, Tipton, and Kiewlicz (1966) show that 56 percent have only one host species.

The Oestridae (Order Diptera), or bot and warble flies, utilize mammals as hosts. In the subfamily Hypoderminae larvae burrow into host tissues, and in the Oestrinae they occupy nasopharyngeal or orbital cavities of their hosts. The host-parasite list for Oestridae in the Old World (Zumpt, 1965) shows that 49 percent are specific to a single host species when wild and domestic animals and man are grouped as hosts. Females are strong fliers and lay eggs on hosts, and specificity should not perhaps be expected to be so well developed. Specialization in the center of their geographic distribution is seen where noses offer an unusual site for colonization: *Pharyngobolus africanus* is specific to the African elephant *Loxodonta africana; Rhinoestrus hippopotami* is specific to the hippopotamus *Hippopotamus amphibius*, and *R. giraffae* is specific to the giraffe *Giraffa camelopardalis*. These are the only oestrids found in the nasal passages of the three hosts. Where noses are more similar, a parasite such as *Oestrus aureoargentatus* may attack seven hosts including roan and sable antelope and common and Lichtenstein's hartebeest. Indeed, when domestic animals are available, they are used by several species of oestrid suggesting that these parasites can be quite catholic in their host utilization if the opportunity is present. A significant proportion of cases where only one host is utilized result from the absence of potential alternate hosts. Specificity becomes more common with distance from the center of diversity in east central Africa of the order Artiodactyla, members of which serve as major hosts for oestrids. *Cephenemyia trompe* and *Oedomagena tarandi* parasitize only the reindeer *Rangifer tarandus* in the Old World which is the only deer of the northern tundra. Thus host specificity in Oestridae results partially from limitation in the number of hosts available rather than strong selection pressures for specialization on one or a narrow range of hosts.

Fleas in the family Hystrichopsyllidae (Order Siphonap-

121

tera) are less specific than the Oestridae, based on an analysis of host records in Hopkins and Rothschild (1962, 1966) for the better sampled areas of the Holarctic Realm (European and Mediterranean Subregions of the Palaearctic and Nearctic). Small mammals are utilized as hosts. Adults are wingless ectoparasites while the larvae are free-living in nesting material. Adults are long-lived, even when off a host, and quite mobile because of their jumping ability. When a mobile adult is the parasitic stage, transfer from one host to another, and frequent chance encounters with novel hosts, often results in a species being found on a wide spectrum of hosts. Whether all hosts enable reproductive competence in the parasites is usually not well known. In this family 37 percent are host specific, but 9 percent utilize more than nine hosts and two species have been collected from over 20 hosts.

The trend seen in the fleas is also evident to a greater extent in the louse flies of the family Hippoboscidae (Order Diptera) that utilize mostly birds, but some mammals as hosts. Most species have winged females, at least until a host is discovered; like the streblids, females incubate larvae until fully grown when they emerge to pupate, but unlike the bat flies, larvae usually drop to the ground for pupation. The close association between host and parasite is lost. With a highly mobile adult as the parasitic stage and without ecological isolation as in the streblids we should expect these parasites to be more generalized than the other taxa so far discussed. Records of breeding hippoboscids in the New World (Bequaert, 1956) indicate that only 17 percent of species are host specific, and if accidental hosts are included, only 9 percent are specific. Three species (7 percent) are known to breed on over 80 host species, and when accidental hosts are included, over 100 host species are recorded.

Data on host specificity for families representing these five major groups of insect parasites on birds and mammals (Table 5.2) show that differences in the adaptive syndromes of the parasites result in a spectrum of specificity. This specificity is reflected both in the percentage of parasites that utilize only one host and the maximum numbers of hosts utilized by any one parasite species. Therefore as an index of specificity for a parasitic family the percent of species that utilize only one host will be used, termed *percent specificity*. For these five families of parasites estimates are now available for the number of parasite species in the world, the number of available potential hosts in the world, and the percent specificity in each family (Table 5.3).

TABLE 5.3. The size and specificity of insect parasite families on birds and mammals and the number of hosts apparently available for colonization. For comparison the Ichneumonidae, which parasitize mainly lepidopteran caterpillars, are included.

Parasite Family	No. Hosts In World	No. Parasite Species in World	Percent Specificity
Philopteridae	8,600	1,500	87
Streblidae	900	200	56
Oestridae	3,100	100	49
Hystrichopsyllidae	3,100	400	37
Hippoboscidae	8,600	100	17
Ichneumonidae	215,000	14,800	53

A comparison between number of parasites and number of hosts poses some problems. Why are hippoboscids so much less speciose than philopterids when the same number of hosts is available? Why have oestrids and hippoboscids speciated to the same extent when opportunities for adaptive radiation on mammal hosts is less than on bird hosts? How have the streblids speciated to such an extent on relatively few hosts? In this case the regression of the number of parasites on number of hosts is not significant (Figure 5.6) ($r^2 = 0.26$, $p > 0.25$).

If, however, percent specificity and host abundance are used in conjunction to predict the relative number of parasites in each family, the major anomaly of the Hippoboscidae is removed. The less specific the taxon is, the fewer the dis-

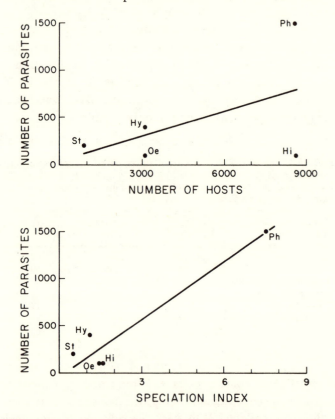

FIGURE 5.6. Number of parasite species in each family in relation to the number of hosts available (above) and the speciation index (below). St indicates Streblidae; Hy, Hystrichopsyllidae; Oe, Oestridae; Ph, Philopteridae; Hi, Hippoboscidae (after Price, 1979).

crete resources for colonization and the slower the rate of speciation. Thus for predictive purposes the product of the number of hosts in the world divided by 1000 and percent specificity divided by 100 provides a speciation index that is significantly correlated with the number of parasites per family because the Hipposboscidae has an index identical to the Oestridae (Figure 5.6). Unfortunately the analysis is not compelling because there is only one point with a high speciation index and it is difficult to find other families of parasites on vertebrates that yield intermediate or more extreme speciation indices. Given the crudity of the data utilized, we should not expect such an analysis to distinguish between the small families with moderate percent specificities. Regional analyses, particularly for well-studied areas, would be profitable.

To strengthen the analysis, parasite families that utilize insects could be used, but here the number of hosts available becomes the major determinant of the speciation index. Data on the Ichneumonidae are provided in Table 5.3 for comparison. When these data are included in the correlation between speciation index and number of species in each parasite family, the slope of the relationship is somewhat lower (b without ichneumonids $= 198$, b with ichneumonids $= 129$), but the regression accounts for 99 percent of the variance.

It would be valuable to include parasites on amphibians and fishes in a similar analysis of specificity since these would add appreciably to the range in number of hosts available. Nematodes, cestodes, or trematodes would prove useful since they are found in all vertebrates, but again analysis would have to be limited to a well-studied region since detailed host-parasite lists are unavailable for many parts of the world.

125

SPECIFICITY ON TWO TROPHIC LEVELS

How does specificity on one trophic level influence specificity and adaptive radiation among parasites of the next trophic level?

Janzen (1973b) argues that tropical herbivores should be more specialized than those in temperate regions because in the tropics there is more spatial and chemical heterogeneity of food resources and, with greater predictability in food supply, specialization can proceed to a greater degree in response to interspecific competition. However, Janzen (1973b) found lower herbivore diversity in the lowland tropics than at intermediate elevations on tropical mountains that suggests many resources are too rare per unit area to support herbivore specialists. The rarity of resources becomes even more extreme for the parasites of these herbivores, and we should expect that adaptive radiation is severely hampered in tropical latitudes. Janzen (1975) has noted that in Costa Rica bruchid weevils (which may be regarded as parasitoids on seeds) have very low levels of parasitism: 35 percent have no known parasitoids of the larval stages and another 38 percent have levels of parasitism well below 1 percent. Janzen argues that bruchids occur in a wide variety of seed chemistries, pod morphologies, fruiting phenologies, microhabitats, seed morphologies, and developmental times, so it is difficult for a parasite to be adapted to a broad enough spectrum of hosts to provide resources per unit area sufficient to maintain a parasite population.

Reduced diversity of parasites of herbivores in the tropics is also illustrated by samples taken in the same way at different latitudes by Owen and Owen (1974) and Janzen and Pond (1975). In both studies more species of parasitic Hymenoptera were found in temperate regions than in the tropics for comparable vegetation types. Rarity of certain resources in the tropics (Janzen, 1973b, 1975) and more intensive predation and disease (Rathcke and Price, 1976; Janzen, 1975) are current working hypotheses on this latitudinal trend.

A comparative study of a group of temperate parasitic herbivores and their hymenopterous parasites supports Janzen's arguments. The Agromyzidae in Britain are used again because a remarkable set of host records for alysiine braconids on agromyzids is provided by Griffiths (1964-1968). Not only are the agromyzid hosts recorded, but the host plants of the leaf-mining parasite from which the braconids were reared are also given. The latter information provides a clue on how parasites adapt to exploit highly specialized hosts.

Many agromyzids are found on only one host plant in the temperate British Isles (Table 5.4). This raises the questions of whether such specialization makes resources too rare for parasites of agromyzids to exploit, and whether the braconids have evolved a more specialized or generalized host utilization pattern to exploit such coarse-grained hosts. The answers are provided in Table 5.4 showing that the braconids are no less specialized than their agromyzid hosts and thus suggesting that adaptive radiation has not been impeded. In general, the braconids are slightly less specific to the plant species they are associated with than their agromyzid hosts (Table 5.4). Thus their area of search in terms of number of plants searched for agromyzids seems to be slightly greater than their hosts, as expected for a parasite exploiting rare resources (see also Janzen, 1973b). Apparently this does not lead to less specificity in braconids than in agromyzids because the braconids have tended to colonize more frequently those agromyzids with numerous hosts. Agromyzids with only one or two hosts represent 77

TABLE 5.4. Percentage of agromyzid parasites in each host number class (from Spencer, 1972), percentage of alysiine braconids that attack agromyzids in each host number class (from Griffiths, 1964-1968), and percentage of braconids in classes of number of plants with which they are associated (from Griffiths, 1964-1968).

Number of hosts	Agromyzids	Alysiine braconids on agromyzids	Alysiine braconids on host plants
1	57	60	53
2	20	14	14
3	7	10	7
4	6	7	6
5	4	2	5
6	3	1	3
7	<1	2	2
8	<1	<1	3
9	<1	1	2
10-19	2	2	4
20-29	<1		<1
30-39			
40-49			
50-59		<1	
60-69			
70-79			<1
Number of Species	267	214	198

percent of all agromyzids in Britain (Table 5.4), but they represent 91 percent (99 of 109 species) of agromyzids with no recorded alysiine braconid parasites (Table 5.5). It does seem that highly specialized hosts are more difficult to exploit than those that are more generalized. The extent of adaptive radiation among parasites of already specialized parasites may well be limited by this difficulty. As Darwin (1872) pointed out, there are serious constraints on adaptive radiation as resources become rare.

Many interesting problems are also posed by the content

TABLE 5.5. Number of agromyzid parasites in each host number class (from Spencer, 1972), and the number of agromyzid species in each class of the number of braconid species on agromyzid species (from Griffiths, 1964-1968) (e.g. 73 agromyzid species with one host species have 0 braconid parasites; 1 agromyzid species with 6 host species has 11 braconid parasites).

Number of hosts	Agromyzids	Braconid parasite number classes											
		0	1	2	3	4	5	6	7	8	9	10	11
		Number of agromyzid species											
1	151	73	35	21	16	3	3						
2	53	26	17	6	2	2							
3	18	2	6	3	3	3	1						
4	15	1	6	4	3	1							
5	11	2	1	1	3	2	1			1			
6	7	1	3		2								1
7	1	1											
8	2		1				1						
9	2	1			1								
10-19	6	2			2			1	1				
20-29	1				1								
Number of species	267	109	69	35	32	12	6	1	1	0	1	0	1

of Table 5.5. For example, what are the properties of agromyzid species that attack many hosts, or their host plants, that make them resistant to attack by parasites? Two species with over ten hosts have no recorded parasites. Conversely, why do some agromyzids with only one host plant support up to five parasite species when almost half the species with one host have no recorded parasites? Why do *Cerodontha pygmaea* (six hosts, eleven parasites) and *Phytomyza ranunculi* (five hosts, nine parasites) have such exceptionally high numbers of associated parasite species? Detailed examination of such questions, preferably using field studies, would be profitable.

It appears that similar processes are operating on parasites of herbivores in tropical and temperate regions; they

129

differ only in degree. In the tropics resources are on the average rare per unit area and therefore adaptive radiation is more limited. Is this rarity induced by greater specificity of the herbivores as suggested by theoretical considerations (e.g. Janzen, 1973b; Dobzhansky, 1950; Williams, 1964; Klopfer, 1959; Klopfer and MacArthur, 1960) or by lower densities of resources without any change in specificity of the parasitic herbivores? A comparison of the number of hosts utilized by tropical and temperate butterflies suggests that tropical species are no more specialized than those in temperate regions (Table 5.6). Only in the case of *Heliconius* species have the host plants been studied in detail so these data afford the best comparison with temperate records. In the other tropical examples specificity is certainly overestimated since many host records have been given only as genera, and for the Malay Peninsula examples of host plants are given as often as complete lists. In cases where it is known that a butterfly uses more than one host but the actual number of hosts is unknown, the records have been placed in a separate category. We must conclude that there is no support contained in Table 5.6 for greater specificity of tropical butterflies and extrapolating from this that the major limitation on radiation of parasites of herbivores in the tropics is reduced density of resources. (Scriber [1973] found evidence for greater specialization in tropical Papilionidae than in temperate members of the family, but he used the number of host plant families to compare species rather than number of host plant species as in Table 5.6.)

Some tentative conclusions have been reached in this chapter concerning the major questions asked at its beginning. The potential for further adaptive radiation in parasitic taxa is very great. Of the resources provided by a typical host taxon relatively few are utilized by parasites in

TABLE 5.6. Specificity of some tropical and temperate butterflies, showing the percentage of species with known food plants that fall into each class of number of hosts utilized.

Number of hosts	TROPICAL			TEMPERATE	
	Heliconius spp. (1)* Mid and S. America	Malay Peninsula (2)	East Africa (3)	North America (4)	Britain (5)
1	24	26	31	22	21
2	14	22	31	21	24
3	16	15	11	12	13
4	4	5	5	11	4
5	6	2	3	7	16
6	2		2	3	9
7	2	1	1	7	3
8	4			1	4
9	2		<1	1	
10-19	12		1	12	4
20-29	8			1	
30-39	8			2	
Unknown but More than one		30	15		
Number of species	51	144	292	214	67

* Sources: (1) Benson, Brown, and Gilbert, 1975; (2) Corbet and Pendlebury, 1956; (3) van Someren, 1974; (4) Tietz, 1972; (5) Stokoe and Stovin, 1944.

a community or throughout the geographic range of the host species. Specificity in host utilization greatly increases the extent of adaptive radiation or its potential. Highly specialized hosts tend to have a low probability of colonization by parasites largely because they provide a relatively rare resource per unit area. Therefore, a taxon of very specialized hosts may dampen the extent of radiation among its parasites more than a moderately specialized group of hosts (but see Evolutionary Concept 2(iv) on comparison between specialized verses unspecialized hosts, the degree of specialization is critical).

SPECIALISTS, MUTUALISM, AND ADAPTIVE RADIATION

The extensive adaptive radiation of some groups of parasites is the more impressive because adaptation has been to food sources that are chemically refractive or so pure as to offer an inadequate diet for a specialist. Cellulose, in plant tissues, plant sap, skin, hair, feathers, and blood all provide grave nutritional problems for would-be parasites, and yet radiation onto these resources has been spectacular. A very significant component in this adaptive radiation has been the acquisition of mutualists that either supplement the parasite's diet with nutrients absent in its food, or they play a role in digestion, or both (Buchner, 1965). As Buchner says, these mutualisms involve "extraordinarily widespread and often fantastically complex bacterial and fungal symbiosis" (1965:3).

Buchner was particularly interested in mutualistic microorganisms in insects, and his and Steinhaus's (1946) extensive summaries of the literature show that of those families listed in Table 2.1 such mutualisms are probably to be found commonly in the Curculionidae, Aphididae, Cicadellidae, Miridae and Philopteridae, in a few Chrysomelidae, and they have also been found in Ichneumonidae (Cowdry, 1923) and Braconidae (Stoltz, Vinson, and MacKinnon, 1976). The importance of such mutualisms in the evolutionary biology of these parasitic groups should not be underrated, and much more study is warranted.

Mutualism as a central force in adaptive radiation is clearly illustrated by the similarity within a taxonomic group in location and form of mycetomes that harbor symbionts, symbiont species, and mode of transmission from

132

parent to offspring. All whiteflies (Aleurodidae) have small, paired, roundish mycetomes in a characteristic position that changes with age. Several common types of mycetome are recognized in the weevils (Curculionidae). Mycetomes in sucking lice (Anoplura) are usually unpaired, roundish, and located below the midgut (Buchner, 1965). Closely related ambrosia beetles cultivate closely related fungi (see Steinhaus, 1946). These symbioses have not resulted from the haphazard acquisition of microorganisms by individual species, although several colonization events per taxon may have occurred in some cases, and many kinds of coexistence are known from some large taxa such as the Cuculionidae.

Given that such "living fossils" as the Peloridiidae (Homoptera) have mutualistic microorganisms, we may well wonder to what extent major taxa such as orders have arisen through colonization of new adaptive zones by the acquisition of mutualists. Also, to what extent has adaptive radiation of microorganisms been dependent upon these mutualistic relationships?

CHAPTER SIX

Ecological Niches, Species Packing, and Community Organization

Ecological Concept 3 and Chapter 3 stress the commonness of non-equilibrium conditions for parasite populations, and Chapter 5 the open, noninteractive nature of parasite communities. The parasitological literature therefore offers a serious challenge to these contentions, since it is full of examples of niche segregation among coexisting parasites, presumably in response to competition, and apparently tight species packing. After an authoritative review of the literature, Holmes (1973:344) concluded that parasite communities "are not 'young' or 'pioneering' ones, but mature communities whose diversity has been established to an important extent through biotic interactions."

In this chapter a close look at species packing in parasite communities will be undertaken in an attempt to resolve the divergence of opinion. Attention is limited to the literature on helminth parasites of vertebrates because it illustrates the full range of types of coexistence seen in any taxon and it contains some of the most celebrated examples of coexistence of parasites, some incredibly rich communities within one organ system, excellent experimental work revealing the fundamental and realized niches (Hutchinson, 1957) of parasites and is deserving of much more attention by ecologists. In the review by Schoener (1974) on resource partitioning in ecological communities covering 80 studies, none on helminths is included.

Some studies are summarized in Table 6.1 with a statement on the predominant type of coexistence in the community. The kind of evidence provided in the studies is also listed with a check indicating quantitative data or experimental manipulation.

The general pattern of coexistence is clear (Table 6.1). The vast majority of cases indicates a predominantly noninteractive niche occupation (no substantial shift from fundamental to realized niche in presence of another species), either in nonoverlapping niches or in strongly overlapping niches. In the latter case all species occur in the alimentary canal and are usually members of very rich helminth faunas. Although studies have not gone far in determining how these species coexist while overlapping so broadly along the length of the alimentary canal, it is clear that other niche dimensions must be considered. Schad (1963a,b) found in the genus *Tachygonetria* in the tortoise gut that some species occur near the gut mucosa while others can be found throughout the gut lumen. Some species feed on large particle sizes, others specialize on bacteria, and others imbibe purely liquid food. Given the great physicochemical diversity of the gut, many other parameters may be involved in the niche diversification in these gut parasites (see MacKenzie and Gibson, 1970; Williams, McVicar, and Ralph, 1970). Thus the group of studies showing predominantly overlapping but noninteractive niche occupation illustrate overlap only on the most obvious dimension of gut length, which has been the most commonly studied. Since niche divergence on other gradients is evident, this group of studies should be classed with the other noninteractive examples making 18 studies out of the total of this type. In the nonoverlapping class some studies have been placed that show seemingly insignificant amounts of overlap (30 percent or less), again making the distinction an arbitrary one.

135

TABLE 6.1. Some studies illustrating coexistence of helminth parasites in vertebrates.
* Indicates presence of an acanthocephalan in the community.

Genus of Parasite	Number of Species	Host and Organs Infected	Reference	Quant. data	Expermtl. manip.	Type of coexistence
1. *Strongyloides*	2	Rat, alimentary canal	Wertheim (1970)	✓	✓	Non overlapping, interactive
2. *Calycotyle*	2	Weasel, proctodeal region	Euzet & Williams (1960)	—	—	"
3. *Aporocotyle* & *Psettarium*	2	Rockfish spp., heart and bronchial system	Holmes (1971)	✓	—	"
4. *Kalicephalus*	3	Racer (snake) alimentary canal	Schad (1956, 1962b)	—	—	"
5. *Filaroides* & others	3	Mink, lung	Stockdale (1970)	—	—	"
6. *Apocreadium*	3	Ocean tally (fish) alimentary canal	Sogandares-Bernal (1959)	—	—	"
7. *Castroia*	2	Bat, alimentary canal	Martin (1969)	✓	—	"
8. *Proteocephalus* & *Neoechinorhynchus**	2	Cisco (fish) alimentary canal	Cross (1934)	✓	—	Non overlapping, Interactive
9. *Tachygonetria*	8	Tortoise, alimentary canal	Schad (1962a, 1963a,b)	✓	—	Overlapping, Non interactive
10. *Tachygonetria* & others	14	Tortoise, alimentary canal	Petter (1966)	✓	—	"
11. *Cylicocyclus* & other strongylids	32	Horse, alimentary canal	Foster (1936)	✓	—	"
12. *Hymenolepis* & others*	43	Scaup, alimentary canal	Hair & Holmes (1975)	✓	—	"

13. *Crepidostomum* & others*	6	Trout, most in alimentary canal	Thomas (1964)	✓	—	"
14. *Cucullanus* & others*	11	Flounder, alimentary canal	MacKenzie & Gibson (1970)	✓	—	"
15. *Capillaria* & others	9	Ray, alimentary canal	Williams, McVicar, & Ralph (1970)	—	—	"
16. *Lepidapedon* & others	15	Cod, alimentary canal	Williams, McVicar, & Ralph (1970)	—	—	"
17. *Hymenolepis* & *Moniliformes**	2	Hamster, alimentary canal	Holmes (1962b)	✓	✓	"
18. *Echinorhynchus**	>1	Fish spp., alimentary canal	Chubb (1964)	—	—	"
19. *Itygonimus* & *Omphalometra*	3+	Mole, alimentary canal	Frankland (1959)	✓	—	"
20. *Hymenoplepis* & *Moniliformes**	2	Rat, alimentary canal	Holmes (1961, 1962a)	✓	✓	Overlapping, Interactive, niche shifts
21. *Proteocephalus* & *Neoechinorhynchus**	2	Stickleback, alimentary canal	Chappell (1969)	✓	—	"
22. *Dactylogyrus*	4	Carp, gills	Paperna (1964)	✓	✓	"exclusion

Holmes (1973) also found many examples of what he called *selective site segregation of niches* where niche occupation was uninfluenced by the presence of other parasite species. As in Table 6.1 he supplied many examples of niche segregation of this type and only few where coexistence involved reduced realized niches in the presence of competitors, i.e. interactive site segregation. He concluded that the majority of parasite communities are mature because the residents have evolved discrete niches without competition. That is, communities have reached the evolutionary phase in Wilson's (1969) concept of community development and have presumably passed the earlier noninteractive, interactive, and assortative phases.

The fact that there are two phases in community development with predominantly noninteractive niche occupation, and these at the two extremes of community age, should make us pause before reaching conclusions from the observation that many parasite communities are predominantly noninteractive. The concepts stressed in this monograph favor the conclusion that parasite communities are in an early stage of development. Noninteractive niche exploitation is common because resources remain unutilized since species have not evolved to use them. Communities are largely in the noninteractive and interactive phases of development and seldom have the assortative and evolutionary phases been reached. Each species is specialized to a narrow range of resources because of coevolutionary demands other than competition. A closer look at the helminth community literature, which follows in the next two sections of this chapter, justifies the latter conclusion in my opinion.

Of course, competition can be observed among helminth gut parasites (e.g. see studies 20 and 21 in Table 6.1), but I contend that it affects a minority of species. The influence of immune responses may also prove to be important, but, again, evidence is wanting that large numbers of species interact via the host's immune system. Crompton (1973) reviewed the literature on 252 species of helminth in the alimentary canal of vertebrates, but he cited no clear examples of interspecific interaction through immunological responses and only four examples of niche shifts through interspecific competition (including Holmes, 1961; and Chappell, 1969 in Table 6.1). But eight examples were provided for the extension of niche breadth in response to intraspecific competition as parasite populations increased in a host, suggesting that interspecific constraints on niche expansion were absent. Particularly in the gut, but perhaps elsewhere also, resources may be replenished by a host so rapidly that they are not limiting except at the most extreme parasite densities (when the host is likely to die with the consequent death of the parasites).

NONINTERACTIVE COEXISTENCE

There is little evidence to suggest that competition has been an organizing force in the studies indicating nonoverlapping, noninteractive coexistence of parasites (studies 1 to 7 in Table 6.1). In mink pulmonary tissues, Stockdale (1970) reports three species of metastrongyle nematode each with a very distinct niche: *Perostrongylus pridhami* in alveolar ducts and terminal and respiratory bronchioles; *Filaroides martis* in peribronchial connective tissue; *Crenosoma hermani* in bronchi. Stockdale notes that pulmonary tissues provide a diverse array of resources available to nematodes: alveoli and alveolar ducts, terminal and respiratory bronchioles, bronchi, lamina propria of the bronchial bifurcation, peribronchial connective tissue and pulmonary arteries. In a range of animals all these sites have been

139

colonized but in the mink little more than three of the six sites are utilized. The nematodes are loosely packed, almost twice as many species could coexist in the pulmonary community, and it is most unlikely that competition could cause such complete divergence of ecological niches.

Two species of blood flukes in rockfishes had fundamental niches that showed a similarity of 7 percent and when together this similarity was reduced by only 3 percent (see data in Holmes, 1971). Indeed, the species could not overlap significantly because they are adapted in fundamentally different ways to lodge in different parts of the vascular system. *Aporocotyle macfarlani* is a squat fluke that wedges into the walls of blood vessels, largely in the gill arches. *Psettarium sebastodorum* is a longer, narrower fluke that loops its body and is so held largely in the chambers of the heart. These species have a grossly different body form that preadapted them for coexistence. Divergence of niches once both species had colonized that host was probably miniuscule. Similar examples are described by Llewellyn (1956), Williams (1960), and Uglem and Beck (1972).

In the snake *Coluber constrictor* three species of *Kalicephalus* can be found: *K. inermis* in the anterior part of the esophagus, *K. costatus* in the duodenum, and *K. rectiphilus* in the rectum (Schad, 1956, 1962b). Species packing is so loose it could not be created by competitive interaction but rather by the demands for specialization, leaving resources available for other species to colonize.

Three species of *Apocreadium* occupy largely nonoverlapping niches: *A. balistis* the first third of the intestine, *A. uroproctoferum* the second third, and *A. coili* the rectum. Had competition been a force in organization we should expect more overlap even to the extent that some species have Hutchinsonian niche differences of only 20 to 30 percent (see Hutchinson, 1959).

In the literature describing species with largely overlapping niches the evidence for competition commonly acting as an organizing force is also very unimpressive. Schad (1963b) claimed that the pinworm communities in tortoises showed nonoverlapping niche occupation with the inference of tight species packing. But this claim was based on correspondence of relative abundances to MacArthur's (1957) broken stick model that can be derived from very different assumptions and is thus discredited as a means of distinguishing organizational influences in a community (Pielou, 1969; Poole, 1974). Based on Schad's (1962a, 1963a, b) studies, we could reasonably expect in the tortoise gut a group of species specialized to each of at least four sections of the gut, each group segregated into those in the lumen and those mostly against the mucosa, and within these subgroups species feeding on large and small particles and liquids. Even at this moderate scale of specialization 24 species could be expected in the gut whereas Schad, Kuntz, and Wells (1960) found only 10 species of *Tachygonetria*. Holmes (1973) chose to emphasize the complementary distributions of *Atractis dactyluris* with *Mehdiella uncinata* and *Tachygonetria dentata* found by Petter (1966) in tortoise guts. But of the 14 species of common nematode studied in detail by Petter, *A. dactyluris* was the only species possibly involved in competitive interactions, and it became dominant only in older tortoises when many other factors could be changing in the gut.

Of the 43 species of helminth in the scaup alimentary canal, Hair and Holmes (1975:253) concluded that the community is "composed of a chance combination of ecological specialists." Where they claimed to find interspecific competition, only a pair of species was involved and the mean overlap between them was only 14 percent, but in some cases large numbers of each species coexisted over 20 to 30 percent of the intestine, suggesting that competition was not the mechanism leading to the small mean overlap value. In one gut *Hymenolepis abortiva* was abundant but completely overlapped by *H. spinocirrosa*. For the 13 parasite species in one scaup gut (data for which were given in detail), there was a significant correlation between population size and the number of segments of the gut occupied ($r^2 = 0.26$, $p < 0.10$) indicating niche expansion under population pressure unimpeded by other species. The abundant species were not competitive dominants as there was no sign of complementary abundances in the 10 scaup guts examined, even between *H. abortiva* and *H. spinocirrosa* (which showed a positive but nonsignificant correlation in abundance). Hair (1975) concluded that the majority of helminth species in the scaup responded individualistically to gradients in the gut, which I interpret as indicating a lack of positive or negative relationships between parasite species.

Chubb (1964) noted no interactions between *Echino-rhynchus clavula* and any other parasitic helminths in the guts of fish. Thomas (1964) also concluded after a very extensive study of brown trout parasites that competition between parasites occurs only rarely. In pairwise partial correlations between six helminths over four years and twice a year (165 pairwise comparisons), he found many more positive than negative significant correlations. In a similar set of correlations involving only the four helminths in the alimentary canal whose distributions in the gut largely overlapped, but with up to three seasonal samples per year (264 pairwise comparisons) he found only 14 cases that were significantly and negatively correlated (whereas 13 such cases should be expected by chance) and 12 cases showed significant and positive correlations.

Although the three intestinal trematodes of the mole show some differences in distribution along the gut, the probability of colonization for each species was so low that they hardly ever interacted (Frankland, 1959). Less than one percent of the moles studied had more than one species of the three, and Frankland noted that the distribution of each species, even within an area where it occurs, is very patchy and local. He also found that the presence of these flukes had no effect on the other intestinal helminth parasites of the host.

INTERACTIVE COEXISTENCE

Interactive niche occupation is defined as the substantial shift by one species from fundamental to realized niche in the presence of another species. The cases in Table 6.1 involving interactive coexistence concern relatively few species: three studies on pairs of species (Cross, 1934; Holmes, 1961, 1962a; Chappell, 1969) and one study reporting coexistence of three species of gill parasite but their exclusion by a fourth (Paperna, 1964). Holmes (1973) provides other examples, particularly of competitive exclusion, but the examples in the table are sufficient to illustrate the types of interactive coexistence to be found and the questions raised by such studies.

Three of the four studies showing interactive coexistence involve acanthocephalans (studies 8, 20, 21). In these parasites all nutrients are absorbed through the integument, as in tapeworms, and carbohydrate shortage can be severely limiting (Noble and Noble, 1976). Therefore, these species may be particularly susceptible to exploitative competition for soluble nutrients.

143

The cases studied by Chappell (1969) and Holmes (1961, 1962a) are particularly convincing examples of competition. In the former case the fundamental niches of the two parasites in stickleback showed a percentage similarity (*PS*) of 46 percent, but when occurring together their realized niches had $PS = 17$ percent. In the rat, the two parasites had fundamental niches that were 64 percent similar, but in concurrent infections the realized niches were only 10 percent similar (Holmes, 1961, 1962a).

The excellent experimental studies of Holmes also raise some important questions. The same pair of parasites in rat showed strong niche shifts, particularly by *Hymenolepis diminuta*; but in hamster there was no indication of competition (Holmes, 1962b). Proportional similarities of distributions along the gut were 45 percent in single infestations and 46 percent in concurrent attacks. Also in rat *H. diminuta* had the broader niche on the gut-length dimension while in hamster *M. dubius* had the broader niche. It becomes necessary in order to understand coexistence of parasites and their community organization fully to study the full range of coexisting hosts since parasite species and species pairs behave differently in each host. Also where host tissues are damaged and/or immune responses triggered, infection may be transient and a rapid succession of species may be expected. Niche occupation and community development in time may be extremely dynamic.

The dynamic nature to be expected in some parasite communities is fully realized in the gill parasitic community of carp studied by Paperna (1964). The community is composed of four species of monogenean trematode in the genus *Dactylogyrus: D. anchoratus, D. extensus, D. minutus,* and *D. vastator.*

Dactylogyrus extensus and *D. anchoratus* were the better colonizers because they could be infective throughout the year. Thus if young carp hatched early in the year when water temperatures were cool, they were rapidly colonized by these species, and significant numbers began to build up on the gill surfaces (Figure 6.1). However, as water temperature increased, *D. vastator* became infective and colonization was much more extensive. This species caused gill damage and rapidly made conditions unsuitable for the early colonizers, and they were usually pushed to extinction on the majority of hosts (see Figure 6.1; 50 days after hatching the maximum number of *D. extensus* found on carp was one). They survived only 10 to 20 days in the presence of *D. vastator*.

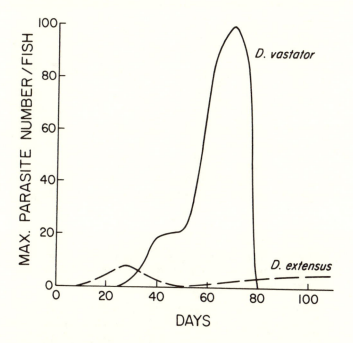

FIGURE 6.1. Maximum number of *Dactylogyrus extensus* and *D. vastator* found per fish in relation to the number of days after fish hatch (after Paperna, 1964).

145

As the gills became seriously damaged by *D. vastator*, conditions on the host became unsuitable for it, and after a relatively brief tenure of 44 to 45 days per host, the populations became extinct.

The gills gradually healed and again became available for colonization by the parasites. But a specific immune response to *D. vastator* eventually developed, the species was eliminated permanently from the community, and the gills were available to the other gill parasites. The very dynamic, non-equilibrium state of the populations and community are also illustrated in Figure 6.2, where fish length is closely correlated with fish age.

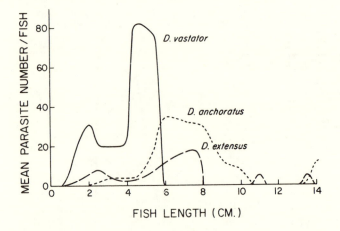

FIGURE 6.2. Mean number of parasites per fish for the species *Dactylogyrus vastator*, *D. anchoratus*, and *D. extensus* in relation to fish length (after Paperna, 1964).

This example contains the important elements of non-equilibrium coexistence within patches identified by Skellam (1951) and Hutchinson (1953) and is similar in some respects to the transient competitive displacement observed by Istock (1967). The competitively inferior species are the better colonizers and appear to have a lower reproductive capacity. New sites are continually becoming available for colonization. Sites rapidly become unsuitable for the competitive dominant, so its competitive edge is repeatedly lost.

Many other examples in the parasitology literature may fit this scenario of transient competitive interactions, but in the absence of good experimental work interpretation of the mechanisms that produce complementary distributions is impossible. For example, *Cucullanus heterochrous* is evenly distributed in the flounder gut from November to May, but from June to October it is found predominantly in the rectum when *C. minutus* is present in the intestine (MacKenzie and Gibson, 1970). As the authors point out, we cannot tell if this is a competitive effect or an age effect on *C. heterochrous*.

Rohde has made a particularly intensive study on species packing in monogenean ectoparasites of fish. He has repeatedly concluded that many resources remain unexploited, that coexistence is noninteractive, and that "segregation of sympatric species may be due to random selection of niches" (Rohde 1977a:164, 1977b, c, 1976, 1978). His evidence is convincing. Monogenean flukes usually exist at low population densities (Rohde, 1977a), and although there is a latitudinal gradient of increasing diversity toward the tropics, there is no change in host specificity along this gradient or specificity in site selection on a host (Rohde, 1976). Monogenean species show highly restricted site selection on hosts even when only one species occurs on a host (Rohde, 1977a). When several species coexist, niche overlap is frequently minimal or nonexistent (Rohde, 1977a,b,c). A comparison of niche occupation on temperate versus tropical hosts yielded the conclusion that most microhabitats utilized by tropical parasites remain empty on fish in cold seas (Rohde, 1976). The mechanisms that result in such narrow niches on a host, even in the absence of competing species, are not fully understood, but the great topographic diversity on fish, particularly on the gill where many monogenean species live, and the considerable complexities of water flow no doubt demand extreme site specificity. This seems to be at least as important as site specificity as a mechanism for increasing population density and the probability of cross-fertilization, an explanation favored by Rohde (1977a).

Community ecology has been dominated by Gause's principle that species cannot coexist for long if they exploit a very similar set of resources. The evolutionary pressure exerted by competition has been regarded as the major organizing influence in communities (e.g. see Schoener, 1974; Cody, 1974b). But for the majority of parasites listed in Table 6.1 competition appears to be uncommon in the present and to have played an insignificant role in their evolutionary history. Where competition exists, it may frequently result in non-equilibrium transient competitive displacement, a disruptive influence in community development. The arguments and data presented in Chapter 3 and 5, based on parasites of plants and animals, fully support these generalizations. For such highly specialized organisms as parasites living in complex environments with steep gradients of resources, even small, randomly generated differences in physiology, behavior, and morphology may result in optimal exploitation of very different sites in a host or very different hosts. Resources available are so diverse it is improbable that a new colonist carries a niche exploitation pattern very similar to a resident species.

CHAPTER SEVEN

Parasite Impact on the Evolutionary Biology of Hosts

It is hard to imagine what the world would be like without parasites. It is even harder to estimate the full impact that parasites have had. Since every species has its set of parasites that act selectively in a nonrandom manner, the repercussions on host biology have been enormous. One of the problems in measuring effects in such tightly coevolved systems is that the moment in evolutionary time that we are able to observe gives us insufficient clues to past events. Nevertheless, even in our primitive state of understanding, the accomplishments of parasites as agents of natural selection on their hosts are impressive and very diverse.

A fair treatment of all aspects of parasite impact on hosts cannot be given in one small chapter. Therefore several important areas will be brushed over leaving room for a little more detail on parasite impact on host population biology and community organization. This area is emphasized because the stage seems to be set for very significant developments, particularly if the several disciplines involved with parasites recognize each others contributions.

Defensive strategies against parasites are, of course, enormously varied. They range from allelochemicals in plants (e.g. Whittaker and Feeny, 1971) including phytoalexins (e.g. Deverall, 1977) and the immune response in vertebrates (e.g. Marchalonis, 1977), through individual be-

149

havior such as defensive displays (e.g. Prop, 1960), grooming (e.g. Collins, 1975, in flies; Struhsaker, 1967, in primates) and preening (e.g. Nelson and Murray, 1971; Kethley and Johnston, 1975), to group defense (e.g. Tostowaryk, 1971; Morris, 1976) and nest design (e.g. Lindquist, 1969), and even to the development of sociality (Michener, 1958; Lin and Michener, 1972) and specialized workers in ant colonies (Eibl-Eibesfeldt and Eibl-Eibesfeldt, 1967). Freeland (1976) describes the enormous impact that pathogens have probably had on the evolution of primate behavior and social organization. Parasites may not have acted as the primary selective agent in all cases, but they played a significant role in the evolution of many host characteristics. The rapidity of host evolution in response to a parasite has been documented for the myxoma virus in rabbits (Fenner and Ratcliffe, 1965).

Parasites may also influence host behavior temporarily. In many cases probability of transmission to another host is greatly increased either by increasing the chances of accidental ingestion or by making the host more vulnerable as prey to a predator involved in the parasite life cycle (e.g. Holmes and Bethel, 1972). But in other cases parasites predispose their host and themselves to other forms of mortality, notably predation, without the benefit of transmission (e.g. Shapiro, 1976; Schaller, 1972; and review by Anderson, 1979).

Parasites have also mediated the evolution of additional symbioses with the host such as mutualistic cleaning (e.g. Limbaugh, 1961), perhaps the evolution of parasitic flowering plants (Atsatt, 1973), and many plant and insect relationships (e.g. Batra and Batra, 1967; Graham, 1967). Mycorrhizal fungi in several cases have been shown to be parasitic on a third species (Björkman, 1960; Campbell, 1968). In an admittedly speculative moment I believe that

150

it is not impossible for parasites to have caused malformations in plants, which so interested van Steenis (1969). If such teratisms were adaptive for the host, the now mutualistic pair of species could perpetuate, and the hopeful monster become established. But I return rapidly from such flights of fancy. Brood parasites have been shown by Smith (1968) to increase host fitness under certain circumstances. Breeding oropendulas in the presence of bot fly adults allow cow birds in their nest since these brood parasites remove immature stages of bot flies by grooming before the oropendula young become infected. While this may seem to be an isolated example, I submit that the isolation stems from our shortcomings in their identification rather than their rarity in nature.

Parasites change hosts in so many ways that it would be surprising if some such changes were not adaptive for the host. The cost paid by the host need be a no greater drain on its energy and nutrient budget than any other tactical maneuvering of metabolic resources in evolutionary time. Mutualists in many instances can be regarded as parasites whose services have been put to good use by the host like the celebrated cases of fig wasps (see Wiebes, 1976 for references) and yucca moths (e.g. Powell and Mackie, 1966).

POPULATIONS AND COMMUNITIES

Among large animals (including man) serious diseases and their impact on populations are sometimes relatively easy to detect because carcases remain visible for a protracted period; numbers before and after an epidemic are roughly known, and behavior of diseased animals may be abnormal. The importance of disease, for example in big game animals and man, should make us wary of any population study on animals or plants, or any theory on population dynamics, that does not include the parasite-disease complex.

Parasites influence every aspect of the population biology of an organism: population size, temporal and spatial dynamics, coexistence, and competition. Some examples follow in order to illustrate the variety of influences parasites may have on hosts and associated organisms. It is often impossible to unravel man's influence on epidemiology in wild animals and plants and therefore, its presence having been acknowledged, the subject will be left largely unexplored. Perhaps man's main function has been to increase the probability of colonization by parasites and therefore the frequency of catastrophic epidemics, particularly where sources of colonists and potential hosts were very distant (see Elton, 1958; Crosby, 1972). Epidemics in this century on three tree species that changed the character of the landscapes they occupied were initiated by man: the chestnut blight on sweet chestnut in North America, the Dutch elm disease on elms in Europe and North America, and the canker on cypress in Italy. Without man's presence the probability of these colonization events would be extremely low, but given enough time such events may still have occurred.

The population dynamics of bighorn sheep in North America are dominated by massive mortality resulting from infection by the lungworms *Protostrongylus stilesi* and *P. rushi* that then predispose the animals to bacterial and viral pathogens causing pneumonia. A foetus can become infected through the mother's placenta and mortality in lambs can be excessive (Hibler, Lange, and Metzger, 1972). All authorities are agreed that the lungworm-pneumonia complex is one of the most important mortality factors in sheep populations (Forrester, 1971; Stelfox, 1971; Uhazy, Holmes, and Stelfox, 1973) and perhaps the most important (Buechner, 1960). Mortalities of 50 percent to 75 percent are reported in these references and even 95 percent in some herds in East Kootenay, British Columbia. Some rangeland became empty after the 1941-1942 epidemic in Banff National Park, south of Bow River (Stelfox, 1971). The spatial dynamics of bighorns was just as much affected as abundances, and before man arrived such drastic shifts in numbers and position must have had complementary impact on many other organisms, most obviously plants on the rangeland and other ungulates. One might almost regard the lungworms as organizer species in early successional communities of western North America.

Where the parasite is less specific and the range of hosts more diverse, the repercussions of an epidemic can be enormous and complex since virulence differs from one species to another, thus radically shifting the relative abundance of species, the ranges they occupy, and the feeding pressure on organisms in their diets. Rinderpest, caused by a virus, has at least 47 natural artiodactyl hosts (Scott, 1970), most of which occur in the center of adaptive radiation in Africa. The disease is usually fatal in buffalo, eland, kudu, and warthog, less frequently so in many species such as bushpig, duiker, giraffe, and wildebeest, and other species such as gazelle, hartebeest, hippopotamus, and impala usually suffer little from the disease. A major pandemic in the late 1800s swept through Africa leaving vast areas uninhabited by certain species and even today distributions reflect the impact (Pearsall, 1954; Spinage, 1962). For example, zebra (Perissodactyla) and some buck species were exterminated by rinderpest in an area now the Queen Elizabeth National Park in Uganda, and when Pearsall (1954) wrote, they had still not recolonized the area. In addition, the pandemic resulted in the disappearance of

153

tsetse flies from large areas of Africa south of the Zambesi River (Stevenson-Hamilton, 1957) and probably the Sudan (Ford, 1970) (see Mulligan, 1970 for additional references). Where the vector is absent, trypanosomiasis cannot exist, resulting in considerable change in human, cattle, and wild animal population distributions and abundance. The repercussions caused by one parasite species can be far-reaching, and one wonders to what extent other major patterns of life have been influenced by parasites in such indirect ways.

The cervids and pronghorn antelope in North America are all potential hosts to the meningeal worm *Parelaphostrongylus tenuis*. This parasite at present occurs throughout most of eastern North America and as far west as the Saskatchewan and Manitoba border in Canada, and Minnesota, Oklahoma, and eastern Texas in the United States (Bindernagel and Anderson, 1972; Anderson and Prestwood, 1979). Its usual host is the whitetail deer *Odocoileus virginianus*, which is tolerant to the infection. Bindernagel and Anderson (1972:1349) state that "there is no apparent barrier to its continued spread westward" through the aspen parkland of Saskatchewan and Alberta and thus into major populations of elk and eventually pronghorn antelope (c.f. distributions in Burt and Grossenheider, 1952). Already the meningeal worm occurs within the range of moose, mule deer, and woodland caribou. All species other than whitetail deer suffer severe neurological disease even with very small numbers of nematodes in the brain (Anderson and Lankester, 1974; Anderson, 1972; Anderson and Prestwood, 1979). The differential pathogenicity of *P. tenuis* makes the whitetail deer a potent competitor wherever it carries the worm. Deleterious effects on other cervids probably involve: (1) direct mortality; (2) increased predation and reduced resistance to adverse climatic conditions; (3) interference with mating; (4) adverse effects on maternal care (R. C. Anderson, personal communication).

As the northern forests were opened up by man, the whitetail deer followed; it came into contact with moose and has since replaced moose as the major cervid in Nova Scotia and Maine and elsewhere on its northern range (Anderson, 1972). A dynamic state seems now to exist between moose in parts of its southern range and deer, with moose occupying refugia on high ground and other forested sites, where they remain ecologically isolated from deer during the critical transmission period in spring, summer, and fall. When young moose leave refugia, they are exposed to infection and usually die, but when deer populations are low, probability of infection declines and moose range can expand. Declines in deer populations may be caused by several mortality factors or ecological succession making the habitat unsuitable (Anderson and Prestwood, 1979). In places where moose are not ecologically isolated from infected deer populations, due to low habitat diversity, they are liable to local extinction.

Whitetail deer have also replaced mule deer and woodland caribou in some parts of their ranges, and reintroduction of caribou into range now occupied by whitetail deer is probably impossible (see Anderson, 1972). Indeed repeated attempts to colonize cervids such as elk, caribou, reindeer, and red deer have been largely unsuccessful in the presence of meningeal worms (Anderson, 1976; Anderson and Prestwood, 1979).

As Anderson (1976) states, the situation surrounding *P. tenuis* is dynamic and complex. Terrestrial molluscs act as intermediate hosts whose distribution and population dynamics are heavily influenced by humidity, availability of which is patchy in time and space. Fires, succession, and human activity open and close habitat for deer and moose.

Age structure in moose populations may be altered when exposed to *P. tenuis* thereby influencing performance of the population for many years into the future.

The biogeography of parasites and their hosts is a vast subject that has been largely ignored by biogeographers. The meningeal worm is now playing a major role in determining the ranges of cervids in eastern North America and its potential for impact in the West is great. Rinderpest must be considered in distributions of ungulates and their parasites in Africa. Whenever a parasite has differential impact on host populations, its influence on distribution and abundance can be profound as the following examples illustrate.

Park (1948) demonstrated the influence of the sporozoan parasite *Adelina tribolii* on flour beetles in single species culture and in competition. When alone, both *Tribolium castaneum* and *T. confusum* cultures in the absence of *Adelina* had about 20 percent more adults in the population than infected cultures. The equilibrium number of beetles in a single species culture was almost three times higher for *T. castaneum* without *Adelina* than in its presence, while population size of *T. confusum* was uninfluenced by the parasite. In the absence of *Adelina* in warm moist conditions, *T. castaneum* usually won in competition with *T. confusum*, the latter becoming extinct, but when present *T. confusum* usually caused the extinction of *T. castaneum*. Clearly the parasite had many and varied influences on the species:

(1) age structure of both species was altered;
(2) the parasite was more deleterious to *T. castaneum* than *T. confusum*;
(3) the equilibrium number of *T. castaneum* in a culture was greatly reduced;

(4) the outcome of competitive encounters was modified, enabling *T. confusum* to expand its niche breadth into warm moist sites at the expense of *T. castaneum*.

Park was obviously fascinated by the impact of *Adelina* and recognized that his early work raised many questions. Unfortunately, although the role of *Adelina* in species interactions is as important and interesting as the study of competition in sterile cultures, the system has never been used again to follow up on Park's lead. If *Adelina tribolii* could be found again, and, given the available genetic stocks of *Tribolium*, we would have an excellent system for potent analytical work on parasite evolutionary biology.

Similar results to those on *Tribolium* were seen in competition between barley and wheat (Burdon and Chilvers, 1977a). Barley was the stronger competitor except in the presence of a specific pathogen, the barley mildew *Erysiphe graminis* f. sp. *hordei*, which reduced prowess making the grasses more or less equally matched. Because the species occupy very similar ecological niches, the authors argue that the parasite is an important factor in coexistence of wheat and barley.

Barbehenn (1969:34) recognized the importance of parasites in mediating interactions between two species and commented on the great "potential significance of the host-parasite competitive mechanism to developing theory in zoogeography, speciation, and in the structure and function of ecosystems." He noted that coexistence of species could be maintained if the weaker competitor carried parasites that were more pathogenic to the stronger, like in *Adelina* and *Erysiphe*, and that anomalous evidence regarding competitive displacement between small mammal species may best be accounted for by differential pathogenicity of a parasite between the competing species. The spectacular

157

exclusion by the Norway rat (*Rattus norvegicus*) of the roof rat (*Rattus rattus*) suggests that epidemic disease in the latter may well have opened up territory to the invader. The mechanism has been well documented for intraspecific competition between Old World man and his conspecifics in the New World (see Crosby, 1972 for an introduction to the literature). The Old World diseases of smallpox, measles, typhus, and chicken pox for which the population carried some immunity, were devastating in Amerindian populations from Alaska to Brazil. The invasion by *Peromyscus maniculatus* of caves occupied by *Neotoma cinerea* is probably facilitated by sylvatic plague (caused by *Yersinia pestis*) to which *Peromyscus* is less susceptible than *Neotoma* (see Nelson and Smith, 1976). In a similar manner, a stomach worm *Graphidium strigosum* apparently enabled rabbits to invade habitats occupied by hares (Broekhuizen and Kemmers, 1976).

As with interactions between races of man, intraspecific competition may be profoundly influenced by parasites. Parasites were strongly implicated in the total failure of reproduction in many populations of the red-spotted newt *Notophthalmus viridescens* suggesting that interdemic selection could play a significant role in the evolution of these newts (Gill, 1978). Those grouse that were unable to establish territories carried a greater load of gastrointestinal parasites than territorial birds (Jenkins, Watson, and Miller, 1963) and were subject to more predation (Jenkins, Watson, and Miller, 1964). In this case, however, it is not clear whether the birds were competitively inferior because of their parasite load or whether they acquired more parasites because of their nonterritorial status.

Barbehenn (1969) found that whereas diversity of rat species was greater on the Malayan mainland than on nearby islands, total rat density was lower. An inverse relationship existed between diversity and total individuals per unit area, thus conflicting with theory on coexistence and species packing (e.g. Darwin, 1872; Dobzhansky, 1950; MacArthur, 1968; Cohen, 1968) and species diversity (e.g. Pianka, 1966), which has not predicted such a result. Although many possible explanations exist, the most parsimonious is that where diversity is high, cross infection of parasites between species is high keeping realized niches much smaller than potential niches, amplifying spatial segregation, and even causing distributional gaps between species. Barbehenn (1969) proposed three conceptual models that may result in the inverse relationship he observed.

Perhaps one of the most pervasive influences of parasites is to maintain distance between species or conspecifics. Any scale may be adequate to minimize cross infection. Such impact in ecological time inevitably results in evolutionary solutions among hosts.

Cornell (1974) examined the case for distributional gaps between bird species where suitable conditions for occupation of the gaps seemed to exist. Current ecological theory provides few insights (e.g. MacArthur, 1972; Cody, 1974a, b). One simple but untested explanation is that gaps are maintained by the capacity of vectors to travel between populations carrying a parasite highly pathogenic to one, but to which the other has adapted. Occupation of the gap by the adapted population is prevented by similar conditions working in the opposite direction.

A probable case of such a phenomenon, described by Warner (1968), concerns the transmission of bird malaria and other pathogens by the mosquito *Culex pipiens fatigans* from migrating shorebirds to indigenous birds in Hawaii. The mosquito was accidentally introduced into Hawaii in 1826. Those indigenous bird species coming within range of infected mosquitoes, being highly susceptible to exotic diseases, were driven to extinction. Only those bird species restricted to mountain forest vegetation have remained unexposed to the introduced diseases. Thus the gap in distribution between migrating shorebirds as the source of disease and indigenous Hawaiian birds is maintained by the mosquito vectors of lethal pathogens.

Interspecific gaps are generated on other dimensions. Where the dispersal phase is the adult and it searches visually for food for its parasitic progeny, visual dissimilarity between related hosts is selected for (Gilbert, 1975). The limited number of discrete shapes of leaf sets limits on the number of host plant species that can coexist in the genus *Passiflora*. Parasitic wasps reach concealed hosts for their larvae with an elongated ovipositor and depth may become the dimension along which hosts are arranged. Many dimensions separate gall wasps in the genus *Cynips* (Askew, 1961), and much of evolutionary maneuvering in communities seems to be under selection for parasite-free space (Lawton, 1978). For example, *Cynips divisa, C. longiventris*, and *C. quercus-folii* coexist on oak in Britain. *Cynips divisa* forms small hard galls, *C. longiventris* medium hard galls, and *C. quercus-folii* large soft galls—all on the undersides of oak leaves. In the sexual generation, *Cynips divisa* f. *verrucosa* formed the larger galls without hairs on young leaves or shoots, while *C. longiventris* f. *substituta* and *C. quercus-folii* f. *taschenbergi* formed smaller pubescent galls in adventitious buds. As a result of these differences, no two species shared the same set of parasitic wasps and inquilines and no parasite was able to exploit both the agamic and sexual generations of the three *Cynips* species. Askew (1961:256) remarks: "If, as seems likely, the populations of these three *Cynips* species are being controlled by their parasites and inquilines, their different parasite and inquiline complements (both generations considered) probably enable them to persist together in the same area, since they are not in complete competition with each other for avoidance of parasites."

The same kinds of selection operate intraspecifically to generate individual distances, low probability of parasite attack, and in effect parasite-free space for the individual (see also Lawton, 1978; Lawton and Schröder, 1977; Otte and Joern, 1976; Zwölfer, 1975). Janzen's (1970) graphical model, with its many variations, of the influence of specific herbivores (mainly fungi and insects) on the distance between conspecific trees can be generalized to encompass the impact of parasites on various host characteristics (Figure 7.1). Distribution of progeny is likely to be similar to I, and the probability of escape from parasite attack (P) will increase to an asymptote with distance from parent or neighbor. Thus the largest number of surviving progeny

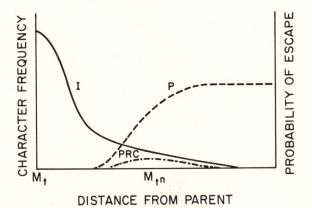

FIGURE 7.1. Janzen's (1970) model generalized to apply to various host characters, where the frequency distribution of a character among progeny (I), relative to that of the parent at the origin, interacts with the probability of escape from parasites (P) to shape the population recruitment curve (PRC). M_t indicates the character mode in the parental population and M_tn a newly selected mode after n generations under the circumstances discussed in the text.

161

is likely to be found where the product of I and P is greatest, as is indicated by the population recruitment curve (PRC). Distance is equally relevant to the impact of seed parasitoids studied by Janzen (e.g. 1970, 1975), the epidemiology of, for example, sylvatic plague in prairie dogs (e.g. Koford, 1958), and the epidemiology of plant diseases (e.g. Burdon and Chilvers, 1975a, b, 1976a, b, c, 1977b). Distance may be measured in space, genetic distance, immunological distance, depth of concealment, shape (for visually hunting organisms), distance between kinds of defense, and so on. In the absence of interspecific influences by parasites selection will be strong to increase the mean and maximum distance between parent and offspring that may be achieved by effective dispersal to achieve spatial distance, outcrossing, and recombination to maximize genetic distance (see also Jaenike, 1978) or the phenotypic plasticity that may influence variety in shape, defense, depth of concealment, and so on. If the parasite is constrained from making modal shifts in characters affecting host colonization in evolutionary time, perhaps because a preferred host is evolutionarily static for certain characters, selection may work to move a host character from the original mode (M_t) closer to a mode coincident with PRC ($M_t n$). Only spatial distance is exempt from this possibility.

In the presence of related hosts with parasites more or less pathogenic to all, a species may be tightly constrained in its evolutionary maneuverability (Figure 7.2). An evolutionary move away from the mode may bring a species into the sphere of infection by new parasites originally harbored by another host. Thus positions of hosts on gradients can be strictly defined by mutual or potentially mutual parasites as suggested by the hypotheses and evidence discussed above (Gilbert, 1975; Askew, 1961; Barbehenn, 1969; Cornell, 1974). But as the mode of one species shifts,

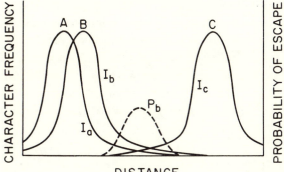

FIGURE 7.2. Frequency distributions (*Ia, Ib, Ic*) of a character that influences parasite attack around the modes of three host species: *A, B, C*. Since species *A* and *B* are very similar in this character, selective pressure exerted by parasites would move the modes apart, assuming that only one dimension is available for such shifts. But the movement of *B* from *A* in evolutionary time is restricted by the pressure of *C* that harbors potential parasites of *B*. This makes P_b (the probability of escape from parasites by species *B*) normally distributed and equidistant from the modes of *A* and *C* if the new parasites are equal in potency to the original parasite community. The mode of the character in *B* would eventually coincide with the recruitment curve defined by I_b and P_b.

it may become adaptive for that of other species to shift. The dance floor that Whittaker (1969) envisages for plants becomes an apt metaphor for the evolutionary interplay between species in space and time moderated by their parasites. Each species dances to keep its individual distance to an evolutionary tune piped by the parasites.

Such a dance may be in progress between the gall wasps *Cynips divisa, C. longiventris,* and *C. quercus-folii* with the tune being played by the parasitic Hymenoptera that oviposit through the gall and into the larval chamber (see Askew, 1961). Only one dimension of the multidimensional dance will be discussed. If *C. divisa* is taken to be the species closest to the ancestral type, then gall thickness has been increased both by stimulating more gall growth centrifugally and by reducing the radius of the larval chamber (Figure 7.3). The strategy has tended to reduce the number of parasites and inquilines associated with each species: *C. divisa* with 14 species, *C. longiventris* with 11 species, and *C. quercus-folii* with 10 species. Strongest selection is presumably operating on *C. longiventris* to increase its distance

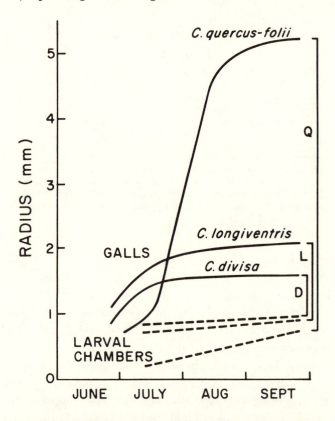

FIGURE 7.3. Development and ultimate size of galls stimulated by *Cynips divisa, C. longiventris,* and *C. quercus-folii.* Wall thickness is indicated on the right (*D, L,* and *Q* respectively) being the difference between gall radius and radius of larval chamber (after Askew, 1961).

from *C. divisa* because these two species in the agamic generation share most parasites. But the further *C. longiventris* galls depart from *C. divisa* galls, the more will *C. longiventris* become vulnerable to parasites harbored by *C. quercus-folii*. As the mode for gall thickness in *C. longiventris* shifts, some parasites may shift or become polymorphic and thus move gradually into the range that enables them to switch to *C. quercus-folii*. Selective pressure will act on *C. quercus-folii* to increase gall thickness or to dance away on another dimension (phenology, location of gall, chemical or mechanical defense of gall). (The relative distances between these *Cynips* species were used to generate Figure 7.2.)

Janzen's model describes the density-dependent response of specific parasites and its effect on spatial arrangement of hosts. Density dependence can also be incorporated into host-parasite models, just as in predator-prey models, in order to examine host abundance in time and its effect on coexisting species. Chilvers and Brittain (1972) developed such a model, which accounted for the coexistence of two species X and Y where in the absence of parasites species X would always win in competition because of a slightly higher growth rate. The model might apply to many coexisting species, but its development was stimulated by the enigmatic coexistence of *Eucalyptus* species in Australia that appeared to occupy almost identical ecological niches (Burdon and Pryor, 1975). For each species of host in the model there was a range of host-specific parasites (e.g. bacteria, fungi, nematodes, insects) grouped as x on host X and y on host Y, which acted in a density-dependent manner. Simple differential equations described the rates of change of the four components. As X increased, the parasites tended to depress the biomass more, leaving resources for Y to coexist; the dominance of Y was prevented by its own specific parasites (Figure 7.4). The model predicts the

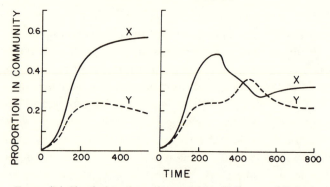

FIGURE 7.4. Simulations from the model by Chilvers and Brittain (1972) indicating the proportions of species X and Y in a community through time (in arbitrary units); (left) in the absence of parasites, with X having faster growth and eventually excluding Y from the community; (right) in the presence of species-specific parasites, when dominance of X is suppressed in a density-dependent manner, permitting coexistence of both hosts. The carrying capacity of the community is 1.

interesting result that the total biomass of X plus Y in the presence of parasites will be higher than in a monoculture of X or Y. This contrasts with Barbehenn's (1969) argument that coexistence of hosts in the presence of effective parasites will lower biomass or numbers compared to that of single species. The difference lies in cross-infection being important in Barbehenn's model whereas it was absent in the Chilvers and Brittain model.

The model has been extended by Burdon and Chilvers (1974a) to include cross-infection by parasites. The non-specific parasites are grouped as pq with proportion p attacking host X and proportion q attacking host Y. If the parasites as a group show preference for host X, even with a faster growth rate, it is pushed to extinction in the presence of Y (Figure 7.5). But if parasites are conditioned by their feeding so they transfer less effectively from one host

166

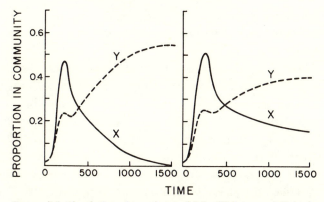

FIGURE 7.5. Simulations from the model by Chilvers and Brittain (1972) modified by Burdon and Chilvers (1974a) to examine the effects of nonspecific parasites on coexisting hosts X and Y through time (in arbitrary units). Parasites can now move freely between hosts and reproduce on both. Parasites that show preference for one host, reproduce more rapidly and cause more damage on that host, and ultimately cause its extinction, even if the host is competitively dominant in the absence of parasites as was species X (left). If, however, parasites become conditioned by feeding on one host and therefore transfer less effectively to the other, coexistence of host species may result (right). In this simulation the bias constant was set at 0.33, meaning that parasites reared on host X could infect host Y with only one third the efficiency of infection on host X, and vice versa. At left the bias constant was 1.0 permitting free movement of parasites between host species and the equal probability of establishing on hosts X and Y.

species to another than within host species, coexistence can be maintained depending on the level of transfer between hosts (Figure 7.5). In the latter case the model can predict lower biomass in association of X and Y than in monocultures of X or Y, but the difference is not as dramatic as observed by Barbehenn. Burdon and Chilvers (1974a) note that in the host conditioning example the parasites produce enough feedback damage to the more abundant host to create equilibrium between hosts, but they are less efficient than the completely host-specific parasites because equilibrium is maintained only within certain parameter limits.

Again, the degree of specificity of parasites becomes important, as seen in Chapter 5, for parasites that differ in specificity seem to differ in their organizing role in host communities. If only one taxon of parasites is studied, another taxon with more or less specific species may counterbalance effects or completely overide them. And yet it is difficult and rare for all taxa of parasites to be studied simultaneously, making interpretation of results risky. The risk occurs when coexistence of parasites (Chapter 6) or their hosts, discussed here, is involved.

Studies on parasites of *Eucalyptus* are to a certain extent reassuring, for insects and fungi on leaves appear to be similar in their specificity and impact and appear not to cancel out the effects of the other taxon. Several common fungi are specific to a single host in associations of two host species and outnumber the nonspecific species (Burdon and Chilvers, 1974a). Insects too were found to be mostly host specific (63 percent of species) in a three species association of *Eucalyptus* (Morrow, 1977a). On an altitudinal gradient over which *Eucalyptus pauciflora* was distributed, the combined effects of fungal and insect damage increased with decreasing altitude (Burdon and Chilvers, 1974b). The relative abundance of *E. pauciflora* and *E. dalrympleana* also correlated nicely with the combined effects of fungi and insects (Burdon and Chilvers, 1974a). In a dense young sapling stand the ratio of *E. pauciflora* to *E. dalrympleana* was 1 to 1.12 but the ratio of leaf area loss was approximately 1:2 (15.1 percent versus 29.9 percent). In a mature stand the ratio of the two species was 1 to 0.38, *E. dalrympleana* having declined dramatically in relative abundance probably because of the greater parasite load. But at this ratio in the mature stand equilibrium may be close because parasite impact was more equitable (7.3 percent versus 9.1 percent effective leaf area loss). Morrow (1977b) also found that *E. pauciflora* was the dominant species in association with *E. stellulata,* although the latter grew much more rapidly in the absence of insect parasites (see also Morrow and LaMarche, 1978). Number of insect species and individuals were both higher on *E. stellulata* and together they killed 96 percent of the shoots (versus 76 percent on *E. pauciflora*) and 50 percent of leaf area (versus 38 percent on *E. pauciflora*). The impact of sap-sucking insects was not measured, but again numbers were much greater on *E. stellulata*. Indeed the cost to *E. pauciflora* of being associated with *E. stellulata* may be considerable since nonspecific parasites near their carrying capacity on the latter may overflow onto the less acceptable host (Morrow, 1977b). Any competitive mechanism that would exclude a congener, such as allelochemicals, may therefore carry a double advantage.

Parasites obviously play a major role in organizing *Eucalyptus* communities, and further studies will no doubt reveal more of the subtlety of the mechanisms involved, and a greater diversity of influential parasites. No mention in this chapter has been made of the seedling parasites *Ceuthospora innumera* and *Piggotia substellata* (see Ashton and Macauley, 1972) that may kill over 50 percent of seedlings of *E. regnans* in some years (Evans, 1976). Dieback of apical shoots of *E. regnans* seedlings is caused by fungi and weevils (Ashton, 1975) reducing their competitive edge against associated noneucalypts. The fungus *Phytophthora cinnamomi* that causes root rot in *Eucalyptus* and many other species must play a dramatic role in regulating the locations and association of eucalypts, given the great range in susceptibility to the disease seen in this genus and variation in influence with change in soil type and topography (Marks, Kassaby, and Fagg, 1975).

The potential for parasite impact on host populations and communities has been grossly oversimplified in this chapter. The full extent of the role of parasites will not be realized quickly. Parasites seldom act alone but predispose the host to others; some may act as vectors of other parasites. Such complexes will greatly increase the possibilities for permutations on various interactive themes. The impact of parasites need be only very slight in order to shift the competitive balance of a host (e.g. Simmonds, 1934, on the competitive ability of Lantana). Conditions in which hosts occur need only change slightly to predispose them to epidemics of parasites. The great impact of parasites on human populations and intraspecific competition, of which we are relatively well aware (e.g. Zinsser, 1935; Crosby, 1972; Burnet and White, 1972; Busvine, 1976; McNeill, 1977), should cause concern for our paucity of understanding on the parasites' role in other organisms.

CHAPTER EIGHT
Further Study

Our science is full of biases because our senses, and therefore our minds, are eclectic. We marvel at long-range dispersal of small organisms and write books on the subject, but lack of movement and population viscosity receives little attention. Embedded in the conventional wisdom is the great mobility of diminutive forms. Visually stimulating organisms, the large, the colorful, the active, the aggressive, command our attention while the secretive and insidious remain largely ignored without any regard for their relative importance. We feast our attention on rare species without the same enthusiasm for rare events. The commonplace is filtered out, the unusual registered. We see predators kill, but rarely parasites, and the visual impact is converted subliminally to a ranking in ecological impact.

These kinds of biases run throughout biology. They play an important role in defining areas of concentration and areas of neglect, in defining concepts that are readily accepted and those that cause commotion, or even consternation.

The study of parasites has repeatedly led us into areas of neglect. Their study has not undermined ecological or evolutionary theory but has demanded a broadening of the conceptual framework. A textbook on ecology with chapters such as diversity and instability, non-equilibrium models, the insignificance of competition, organization in communities through symbiosis, would be unconventional, to say the least. However, it might be representative of the

171

ecologically important processes and conditions experienced by the majority of organisms.

We are so used to thinking and theorizing in terms of one-to-one interactions, and yet the probability of such purity ever being found in nature is low. Be it predator and prey, competitors, or organisms responding to abiotic forces, other unsuspected organisms are usually involved that moderate, exacerbate, ameliorate, or at least complicate the interaction. This does not argue that ecological confrontations are intractably complex, but that alternative explanations admitting the existence of other organisms may become more general or economical. For example, hypotheses to explain the adaptive function of drip-tips of leaves in tropical rain forest have been unconvincing (see Richards, 1952, for review). But as a means of closing to potential herbivorous parasites the cavity in a rolled leaf, it is much more effective than alternative designs, a factor that seems not to have entered the debate over the past 80 years.

Perhaps the greatest challenge for the future is to extend the concepts in ecology and evolution, developed largely from an appreciation of larger organisms (i.e. the less representative and the more unusual), to include realistic concepts relevant to the very small and the very specialized. New conceptual themes need development also. The selective pressure exerted by highly specialized organisms on host populations can produce the most unsuspected results. We are not yet sensitive enough to appreciate even a small fraction of them or even the major classes of effects that parasites may have. Host populations may be pushed to a parthenogenetic mode of reproduction (see Stefani's work discussed by White, 1978). Parasites may act as weapons in intraspecific and interspecific biological warfare, in the former case being a potent force in interdemic selection. Many parasites use mutualists as agents of biological warfare against their hosts by introducing pathogenic fungi and bacteria into host tissues (see Buchner, 1965 for details). Diets of hosts may change to include prophylactics or remedials for parasitic infection (e.g. Weigl, 1968; Dogiel, 1964), or host ranges may be limited by the distribution of such dietary ingredients. Host shifts may render a parasite free from others (e.g. once milkweeds were colonized, their toxins no doubt contributed to the antibiotic properties of haemolymph in the large milkweed bug [Frings, Goldberg, and Arentzen, 1948]). There are many ways in which parasites may disqualify themselves from this classification by increasing fitness of hosts (see Owen and Wiegert, 1976; Smith, 1968). Individual, interdemic, and interspecific selection will favor resistant hosts, host demes, or species at the expense of susceptible hosts.

Indeed, in so many cases the unit of selection becomes host plus parasite. A gall former (itself a parasite) relies on the host for protection against parasites and must manipulate the host to optimize defenses. The selective tension between the three parties becomes spectacularly creative. The gall former gains from increased gall size, but the plant can manipulate resource supply to the gall (see Whitham, 1978, 1979). The parasites are indebted to the plant for easing their task, but the plant becomes mutually in debt for the services rendered by the parasites. The gall former is caught between the two trophic levels united, and squeezing out of this jam has resulted in the most remarkable diversity of galling species, gall shapes and sizes, life histories, phenologies, and morphologies.

Another author in this monograph series titled his last chapter "Speculations." It would be impudent for me to do

173

the same, for speculation runs throughout this book. But it is time for a speculative phase in the development of the ecology and evolution of parasites. Tantalizing clues to the ubiquitous influence of parasites exist, but we have hardly an idea of where these clues will eventually lead or how their tracing will expand or modify existing theory in ecology and evolution. Their study promises a rich supply of new ideas, hypotheses, and theories, a refreshing stimulus for detailed observation and experimentation.

Major themes in this book have been non-equilibrium conditions, noninteractive colonization, genetic systems, and the role of parasites in host biology. Of course, all these subjects demand much more detailed study in the future, and I shall not belabor the point except to emphasize the importance of studies on population structure and genetic systems.

Population structure forms a central theme in studying the evolutionary biology of parasites. It is defined by the attributes of the species, its environment, and the evolutionary consequences of their interaction through time. It defines to a large extent the evolutionary potential of the species and its potential for doing mischief among host populations. For no parasitic species do we have an adequate description of population structure. Detailed mapping of resources relevant to a parasite species are essential, as is the knowledge of the extent to which these resources are colonized, the numbers per patch, effective population size, movement between patches, and patch dynamics in space and time. The mobility of individuals and the fitness of mobile individuals must be known. Parasites can frequently be found well removed from host populations. What is their fitness? Is movement from a patch common or rare? Both the very common and the very rare need study and quantification. We are bad at both. The ecology of rare events, an important aspect of life for parasites, is yet to be developed. Long-term studies only can distinguish the common from the rare as Kennedy's (1975b) study of Slapton Ley demonstrates (see Chapter 3), and yet such a study is a rarity in itself.

The great advances in evolutionary biology since 1966 stimulated by the study of isoenzymes by electrophoresis seems almost to have bypassed the parasites. Yet the potential is enormous for critical studies on the correlation between genotypic variability of a host species and its parasites, diversity of host species attacked versus genotypic diversity of a parasite species, population structure of host versus that of the parasites, genetic distances between related hosts versus their parasites, degree of divergence between parasite populations, modes of speciation, and much more. Cytogenetic studies of parasites also seem to have received low priority although, again, the potential rewards are great. The fusion of what is known of genetics with other aspects of population structure enable a description of genetic systems that are central to understanding so many aspects of parasite biology.

The massive chasms in our knowledge of nature may be illustrated with a simple question: How have parasites of *Drosophila*, for example the parasitic Hymenoptera, responded to the extensive adaptive radiation of that genus on the Hawaiian Islands? Similar questions could be asked of any group of organisms and the humbling response must always be that we just do not know.

The evolutionary biology of parasites requires a synthesis of all aspects of their biology and that of their hosts. No detail is insignificant. Likewise, their study demands the dissolving of conventional barriers in biology, both in tax-

175

onomy and levels of organization. This monograph has attempted a treatment of phenomena in cells through to interactions on a geographic scale, the bacteria and fungi
through to trees and mammals. But as a synthesis of the
evolutionary biology of parasites, it is only a beginning.

Bibliography

Alexopoulos, C. J. 1952. *Introductory mycology*. Wiley, New York.

Anderson, R. C. 1972. The ecological relationships of meningeal worm and native cervids in North America. *J. Wildl. Dis. 8*, 304–310.

Anderson, R. C. 1976. Helminths. In *Wildlife diseases*. (Ed. by L. A. Page), pp. 35–43. Plenum, New York.

Anderson, R. C., and M. W. Lankester. 1974. Infectious and parasitic diseases and arthropod pests of moose in North America. *Natur. Can. 101*, 23–50.

Anderson, R. C., and A. K. Prestwood. 1979. Lungworms. In *The diseases of the white-tailed deer*. (Ed. by F. Hayes). In press. U.S. Dept. Interior, U.S. Wild. Serv., Washington, D.C.

Anderson, R. M. 1974. Mathematical models of host-helminth parasite interactions. In *Ecological stability*. (Ed. by M. B. Usher and W. H. Williamson), pp. 43–69. Chapman and Hall, London.

Anderson, R. M. 1979. The influence of parasitic infection on the dynamics of host population growth. In *Population dynamics*. (Ed. by R. M. Anderson, B. D. Turner, and L. R. Taylor). In press. Blackwell Sci. Pub., Oxford.

Anderson, R. M., and R. M. May. 1978. Regulation and stability of host-parasite population interactions. I. Regulatory processes. *J. Anim. Ecol. 47*, 219–247.

Andrewartha, H. G., and L. C. Birch. 1954. *The distribution and abundance of animals*. Univ. of Chicago Press, Chicago.

Arme, C., and R. W. Owen. 1968. Occurrence and pathology of *Ligula intestinalis* infections in British fishes. *J. Parasitol. 54*, 272–280.

Arthur, D. R. 1965. Feeding in ectoparasitic Acari with special reference to ticks. *Adv. Parasitol. 3*, 249–298.

Ashton, D. H. 1975. The seasonal growth of *Eucalyptus regnans* F. Muell. *Aust. J. Bot. 23*, 239–252.

Ashton, D. H., and B. J. Macauley. 1972. Winter leaf spot disease of seedlings of *Eucalyptus regnans* and its relation to forest litter. *Trans. Brit. Mycol. Soc. 58*, 377–386.

Askew, R. R. 1961. On the biology of the inhabitants of oak galls of Cynipidae (Hymenoptera) in Britain. *Trans. Soc. Brit. Entomol. 14*, 237–268.

Askew, R. R. 1968. Considerations on speciation in Chalcidoidea (Hymenoptera). *Evolution 22*, 642–645.

Askew, R. R. 1971. *Parasitic insects*. American Elsevier, New York.

Askew, R. R. 1975. The organization of chalcid-dominated parasitoid communities centered upon endophytic hosts. In *Evolutionary strategies of parasitic insects and mites*. (Ed. by P. W. Price), pp. 130–153. Plenum, New York.

Askew, R. R., and J. M. Ruse. 1974. The biology of some Cecidomyiidae (Diptera) galling the leaves of birch (*Betula*) with special reference to their chalcidoid (Hymenoptera) parasites. *Trans. Royal Entomol. Soc. London 126*, 129–167.

Atchley, W. R. 1977. Biological variability in the parthenogenetic grasshopper *Warramaba virgo* (Key) and its sexual relatives. 1. The eastern Australian populations. *Evolution 31*, 782–799.

Atsatt, P. R. 1970. The population biology of annual

grassland hemiparasites. II. Reproductive patterns in *Orthocarpus. Evolution 24,* 598–612.

Atsatt, P. R. 1973. Parasitic flowering plants: How did they evolve? *Amer. Natur. 107,* 502–510.

Baer, J. G. 1971. *Animal parasites.* Weidenfeld and Nicolson, London.

Baker, H. G., and G. L. Stebbins. 1965. *The genetics of colonizing species.* Academic Press, New York.

Barbehenn, K. R. 1969. Host-parasite relationships and species diversity in mammals: An hypothesis. *Biotropica 1,* 29–35.

Barnes, H. F. 1951. *Gall midges of economic importance.* Vol. 5. *Gall midges on trees.* Crosby Lockwood, London.

Barnes, H. F. 1955. Gall midges reared from acorns and acorn-cups. *Entomol. Mon. Mag. 91,* 86–87.

Barnes, H. F. 1958. Wheat blossom midges on Broadbalk, Rothamsted Experimental Station, 1927–56. *Proc. 10th Int. Congr. Entomol. 3,* 367–374.

Batra, S.W.T., and L. R. Batra. 1967. The fungus gardens of insects. *Sci. Amer. 217* (5), 112–120.

Beaver, R. A. 1977. Bark and ambrosia beetles in tropical forests. *Biotrop. Spec. Pub. No. 2,* 133–147. (Proc. Symp. Forest Pests and Diseases in Southeast Asia, Bogor, Indonesia.)

Beddington, J. R., C. A. Free, and J. H. Lawton. 1976. Concepts of stability and resilience in predator-prey models. *J. Anim. Ecol. 45,* 791–816.

Beddington, J. R., C. A. Free, and J. H. Lawton. 1978. Characteristics of successful natural enemies in models of biological control of insect pests. *Nature 273,* 513–519.

Beddington, J. R., and P. S. Hammond. 1977. On the dynamics of host-parasite-hyperparasite interactions. *J. Anim. Ecol. 46,* 811–821.

Benson, R. B. 1950. An introduction to the natural history of British sawflies. *Trans. Soc. Brit. Entomol. 10,* 45–142.

Benson, W. W., K. S. Brown, Jr., and L. E. Gilbert. 1975. Coevolution of plants and herbivores: Passion flower butterflies. *Evolution 29,* 659–680.

Bequaert, J. C. 1956. The Hippoboscidae or louse-flies (Diptera) of mammals and birds. Part II. Taxonomy, evolution and revision of American genera and species. *Entomol. Amer. 36,* 417–611.

Berlocher, S. H. 1976. *The genetics of speciation in Rhagoletis (Diptera: Tephritidae).* Ph.D. Dissertation. Univ. of Texas.

Berrie, A. D. 1973. Snails, schistosomes and systematics: some problems concerning the genus *Bulinus.* In *Taxonomy and ecology.* (Ed. by V. H. Heywood), pp. 173–188. Academic Press, London.

Bindernagel, J. A., and R. C. Anderson. 1972. Distribution of the meningeal worm in white-tailed deer in Canada. *J. Wildl. Manage. 36,* 1349–1353.

Björkman, E. 1960. *Monotropa hypopitys* L.—an epiparasite on tree roots. *Physiol. Plantarum 13,* 308–327.

Blackman, R. L. 1977. The existence of two species of *Euceraphis* (Homoptera Aphididae) on birch in Western Europe, and a key to European and North American species of the genus. *Syst. Entomol. 2,* 1–8.

Bonner, J. T. 1965. *Size and cycle: An essay on the structure of biology.* Princeton Univ. Press, Princeton, N.J.

Bouček, Z., and R. R. Askew. 1968. *Index of Palearctic Eulophidae (excl. Tetrastichinae).* Index of entomophagous insects no. 3. (Ed. by V. Delucchi and G. Remandière). Le François, Paris.

180

Bretsky, P. W., and D. M. Lorenz. 1970. An essay on genetic-adaptive strategies and mass extinctions. *Geol. Soc. Amer. Bull. 81,* 2249–2456.

Brinck, P. 1979. Population size of ectoparasites of rodents and shrews. In *Proc. Int. Conf. Fleas.* (Ed. by R. Traub). In press. Balkema, Rotterdam.

Brinkerhoff, L. A. 1970. Variation in *Xanthomonas malvacearum* and its relation to control. *Annu. Rev. Phytophathol. 8,* 85–110.

Broekhuizen, S., and R. Kemmers. 1976. The stomach worm, *Graphidium strigosum* (Dujardin) Raillet and Henry, in the European hare, *Lepus europaeus* Pallas. In *Ecology and management of European hare populations.* (Ed. by Z. Pielowski and Z. Pucek), pp. 157-171. Polish Hunt. Assn., Warsaw.

Brown, W. J. 1959. Taxonomic problems with closely related species. *Annu. Rev. Entomol. 4,* 77–98.

Brues, C. T. 1924. The specificity of food-plants in the evolution of phytophagous insects. *Amer. Natur. 58,* 127–144.

Buchner, P. 1965. *Endosymbiosis of animals with plant microorganisms.* Wiley, New York.

Buechner, H. K. 1960. The bighorn sheep in the United States, its past, present, and future. *Wildl. Monogr. 4,* 1–174.

Burdon, J. J., and G. A. Chilvers. 1974a. Fungal and insect parasites contributing to niche differentiation in mixed species stands of eucalypt saplings. *Aust. J. Bot. 22,* 103–114.

Burdon, J. J., and G. A. Chilvers. 1974b. Leaf parasites on altitudinal populations of *Eucalyptus pauciflora* Sieb. ex Spreng. *Aust. J. Bot. 22,* 265–269.

Burdon, J. J., and G. A. Chilvers. 1975a. Epidemiology of damping-off disease (*Pythium irregulare*) in relation to density of *Lepidium sativum* seedlings. *Ann. Appl. Biol. 81*, 135–143.

Burdon, J. J., and G. A. Chilvers. 1975b. A comparison between host density and inoculum density effects on the frequency of primary infection foci in *Pythium*-induced damping-off disease. *Aust. J. Bot. 23*, 899–904.

Burdon, J. J., and G. A. Chilvers. 1976a. The effect of clumped planting patterns on epidemics of damping-off disease in cress seedlings. *Oecologia 23*, 17–29.

Burdon, J. J., and G. A. Chilvers. 1976b. Controlled environment experiments on epidemics of barley mildew in different density host stands. *Oecologia 26*, 61–72.

Burdon, J. J., and G. A. Chilvers. 1976c. Epidemiology of *Pythium*-induced damping-off in mixed species seedling stands. *Ann. Appl. Biol. 82*, 233–240.

Burdon, J. J., and G. A. Chilvers. 1977a. The effect of barley mildew on barley and wheat competition in mixtures. *Aust. J. Bot. 25*, 59–65.

Burdon, J. J. and G. A. Chilvers. 1977b. Controlled environment experiments on epidemic rates of barley mildew in different mixtures of barley and wheat. *Oecologia 28*, 141–146.

Burdon, J. J., and L. D. Pryor. 1975. Interspecific competition between eucalypt seedlings. *Aust. J. Bot. 23*, 225–229.

Burnet, M., and D. O. White. 1972. *Natural history of infectious disease*. 4th ed. Cambridge Univ. Press, London.

Burns, D. P., and L. P. Gibson. 1968. The leaf-mining weevil of yellow-poplar. *Can Entomol. 100*, 421–429.

Burt, W. H., and R. P. Grossenheider. 1952. *A field guide to the mammals*. Houghton Mifflin, Boston.

Bush, G. L. 1969. Sympatric host race formation and speciation in frugivorous flies of the genus *Rhagoletis* (Diptera, Tephritidae). *Evolution 23*, 237–251.

Bush, G. L. 1974. The mechanism of sympatric host race formation in the true fruit flies (Tephritidae). In *Genetic mechanisms of speciation in insects*. (Ed. by M.J.D. White), pp. 3–23. Australia and New Zealand Book Co., Brookvale, N.S.W.

Bush, G. L. 1975a. Sympatric speciation in phytophagous parasitic insects. In *Evolutionary strategies of parasitic insects and mites*. (Ed. by P. W. Price), pp. 187–206. Plenum, New York.

Bush, G. L. 1975b. Modes of animal speciation. *Annu. Rev. Ecol. Syst. 6*, 339–364.

Bush, G. L., S. M. Case, A. C. Wilson, and J. L. Patton. 1977. Rapid speciation and chromosomal evolution in mammals. *Proc. Nat. Acad. Sci. U.S.A. 74*, 3942–3946.

Busvine, J. R. 1976. *Insects, hygiene and history*. Athlone, London.

Campbell, E. O. 1968. An investigation of *Thismia rodwayi* F. Muell and its associated fungus. *Trans. Roy. Soc. New Zealand Bot. 3*, 209–219.

Carson, H. L. 1968. The population flush and its genetic consequences. In *Population biology and evolution*. (Ed. by R. C. Lewontin), pp. 123–137. Syracuse Univ. Press, Syracuse, N.Y.

Carson, H. L. 1975. The genetics of speciation at the diploid level. *Amer. Natur. 109*, 83–92.

Caswell, H. 1978. Predator-mediated coexistence: A nonequilibrium model. *Amer. Natur. 112*, 127–154.

Chappell, L. H. 1969. Competitive exclusion between two intestinal parasites of the three-spined stickleback, *Gasterosteus aculeatus* L. *J. Parasitol. 55*, 775–778.

Chilvers, G. A., and E. G. Brittain. 1972. Plant competition mediated by host-specific parasites–a simple model. *Aust. J. Biol. Sci. 25*, 749–756.

Chubb, J. C. 1964. Occurrence of *Echinorhynchus clavula* (Dujardin, 1845) nec Hamaan, 1892 (Acanthocephala) in the fish of Llyn Tegid (Bala Lake), Merionethshire. *J. Parasitol. 50*, 52–59.

Cifelli, R. 1969. Radiation of Cenozoic planktonic foraminifera. *Syst. Zool. 18*, 154–168.

Clapham, A. R., T. G. Tutin, and E. F. Warburg. 1957. *Flora of the British Isles.* Cambridge Univ. Press, London.

Claridge, M. F., and M. R. Wilson. 1976. Diversity and distribution patterns of some mesophyll-feeding leafhoppers of temperate woodland canopy. *Ecol. Entomol. 1*, 231–250.

Claridge, M. F., and M. R. Wilson. 1978. British insects and trees: A study in island biogeography or insect/plant coevolution? *Amer. Natur. 112*, 451–456.

Clarke, B. 1976. The ecological genetics of host-parasite relationships. In *Genetic aspects of host-parasite relationships.* (Ed. by A.E.R. Taylor and R. Muller), pp. 87–103. *Symp. Brit. Soc. Parasitol. 14.*

Clausen, C. P. 1976. Phoresy among entomophagous insects. *Annu. Rev. Entomol. 21*, 343–368.

Clay, T. 1950. The Mallophaga as an aid to the classification of birds with special reference to the structure of feathers. *Proc. 10th Int. Ornith. Congr.* (Ed. by S. Hörstadius), pp. 207–215. Almquist and Wiksell, Stockholm.

Cleveland, L. R. 1960. Effects of insect hormones on the protozoa of *Cryptocercus* and termites. In *Host influence on parasite physiology.* (Ed. by L. A. Stauber), pp. 5–10. Rutgers Univ. Press, New Brunswick, N.J.

Cody, M. L. 1974a. Optimization in ecology. *Science 183*, 1156–1164.

Cody, M. L. 1974b. *Competition and the structure of bird communities*. Princeton Univ. Press, Princeton, N.J.

Cohen, J. E. 1968. Alternate derivations of a species-abundance relation. *Amer. Natur. 102*, 165–172.

Cohen, J. E. 1976. Schistosomiasis: A human host-parasite system. In *Theoretical ecology Principles and applications* (Ed. by R. M. May), pp. 237–256. W. B. Saunders, Philadelphia.

Coles, J. F., and D. P. Fowler. 1976. Inbreeding in neighboring trees in two white spruce populations. *Silvae Genetica 25*, 29–34.

Colinvaux, P. A. 1973. *Introduction to ecology*. Wiley, New York.

Collins, N. C. 1975. Tactics of host exploitation by a thermophilic water mite. *Misc. Publ. Entomol. Soc. Amer. 9*, 250–254.

Collins, N. C., R. Mitchell, and R. G. Wiegert. 1976. Functional analysis of a thermal spring ecosystem, with an evaluation of the role of consumers. *Ecology 57*, 1221–1232.

Cooper, A. F., Jr., and S. D. Van Gundy. 1971. Senescence, quiescence, and cryptobiosis. In *Plant parasitic nematodes*. (Ed. by B. M. Zuckerman, W. F. Mai, and R. A. Rohde), 2:297–318. Academic Press, New York.

Cope, E. D. 1885. On the evolution of the Vertebrata, progressive and retrogressive. *Amer. Natur. 19*, 140–148, 234–247, 341–353.

Cope, E. D. 1896. *The primary factors of organic evolution*. Open Court, Chicago.

Corbet, A. S., and H. M. Pendlebury. 1956. *The butterflies of the Malay Peninsula*. 2nd ed. rev. Oliver and Boyd, Edinburgh.

Cornell, H. 1974. Parasitism and distributional gaps between allopatric species. *Amer. Natur. 108*, 880–883.

Cornell, H. V., and J. O. Washburn. 1979. Evolution of the richness-area correlation for cynipid gall wasps on oak trees: A comparison of two geographic areas. *Evolution, 33*, 257–274.

Cowdry, E. V. 1923. The distribution of *Rickettsia* in the tissues of insects and arachnids. *J. Exp. Med. 37*, 431–456.

Craddock, E. M. 1974. Chromosomal evolution and speciation in *Didymuria*. In *Genetic mechanisms of speciation in insects*. (Ed. by M.J.D. White), pp. 24–42. Reidel, Dordrecht, Holland.

Cram, E. B. 1927. Bird parasites of the nematode suborders Strongylata. Ascaridata, and Spirurata. *Bull. U. S. Nat. Mus. 140*, 1–465.

Crofton, H. D. 1971. A model of host-parasite relationships. *Parasitology 63*, 343–364.

Crofton, H. D., J. H. Whitlock, and R. A. Glazier. 1965. Ecology and biological plasticity of sheep nematodes. II. Genetic x environmental plasticity in *Haemonchus contortus* (Rud. 1803). *Cornell Vet. 55*, 251–258.

Crompton, D.W.T. 1973. The sites occupied by some parasitic helminths in the alimentary tract of vertebrates. *Biol. Rev. 48*, 27–83.

Crosby, A. W., Jr. 1972. *The Columbian exchange. Biological and cultural consequences of 1492*. Greenwood, Westport, Conn.

Cross, S. X. 1934. A probable case of non-specific immunity between two parasites of ciscoes of the Trout Lake region of northern Wisconsin. *J. Parasitol. 20*, 244–245.

Crozier, R. H. 1970. On the potential for genetic variability in haplodiploidy. *Genetica 41*, 551–556.

Cuellar, O. 1977. Animal parthenogenesis. *Science 197*, 837–843.

Darlington, A. 1974. The galls of oak. In *The British Oak*. (Ed. by M. G. Morris and F. H. Perring), pp. 298–311. Classey, Faringdon, Berks.

Darwin, C. 1872. *The origin of species*. 6th ed. Murray, London.

Das, K. M., and J. H. Whitlock. 1960. Subspeciation in *Haemonchus contortus* (Rudolphi, 1803), Nematoda, Trichostrongyloidea. *Cornell Vet. 50*, 182–197.

Day, P. R. 1974. *Genetics of host-parasite interaction*. Freeman, San Francisco.

Deverall, B. J. 1977. *Defense mechanisms of plants*. Cambridge Univ. Press, Cambridge.

Dobzhansky, T. 1950. Evolution in the tropics. *Amer. Sci. 38*, 209–221.

Dobzhansky, T., F. J. Ayala, G. L. Stebbins, and J. W. Valentine. 1977. *Evolution*. Freeman, San Francisco.

Dogiel, V. A. 1961. Ecology of the parasites of freshwater fishes. In *Parasitology of fishes*. (Ed. by V. A. Dogiel, G. K. Petrushevski, and Y. I. Polyanski), pp. 1–47. Oliver and Boyd, Edinburgh.

Dogiel, V. A. 1964. *General parasitology*. Oliver and Boyd, Edinburgh.

Dowell, R. V. 1977. Biology and intrageneric relationships of *Bathyplectes stenostigma*, a parasite of the alfalfa weevil. *Ann. Entomol. Soc. Amer. 70*, 845–848.

Dritschilo, W., H. Cornell, D. Nafus, and B. O'Connor, 1975. Insular biogeography: Of mice and mites. *Science 190*, 467–469.

Duke, B.O.L. 1972. Behavioural aspects of the life cycle of *Loa*. In *Behavioural aspects of parasite transmission*. (Ed. by E. U. Canning and C. A. Wright), pp. 97–107. Academic Press, London.

187

Dupuis, C. 1949. On the 'late melanism' of the larval stages of Pentatomidae (Hemiptera Heteroptera). *Entomol. Mon. Mag. 85*, 229–230.

Eastop, V. F. 1973. Deductions from the present day host plants of aphids and related insects. In *Insect/plant relationships* (Ed. by H. F. van Emden), pp. 157–178. Symp. Roy. Entomol. Soc. London 6.

Edmunds, G. F., Jr. 1973. Ecology of black pineleaf scale (Homoptera: Diaspididae). *Environ. Entomol. 2*, 765–777.

Edmunds, G. F., Jr., and D. N. Alstad. 1978. Coevolution in insect herbivores and conifers. *Science 199*, 941–945.

Ehrlich, P. R., and P. H. Raven. 1964. Butterflies and plants: A study in coevolution. *Evolution 18*, 586–608.

Eibl-Eibesfeldt, I., and E. Eibl-Eibesfeldt. 1967. Das Parasitenabwehren der Minima-Arbeiterinnen der Blattschneider-Ameise (*Atta cephalotes*). *Z. Tierpsychol. 24*, 278–281.

Eichler, W. 1948. Some rules in ectoparasitism. *Ann. Mag. Natur. Hist. Ser. 12, 1*, 588–598.

El Mofty, M., and J. D. Smyth. 1960. Endocrine control of sexual reproduction in *Opalina ranarum* parasitic in *Rana temporaria*. *Nature 186*, 559.

Elton, C. S. 1927. *Animal ecology*. Sidgwick and Jackson, London.

Elton, C. S. 1958. *The ecology of invasions by animals and plants*. Methuen, London.

Emlen, J. M. 1973. *Ecology: An evolutionary approach*. Addison-Wesley, Reading, Mass.

Endler, J. A. 1977. *Geographic variation, speciation, and clines*. Princeton Univ. Press, Princeton, N.J.

Euzet, L., and H. H. Williams. 1960. A re-description of the trematode *Calicotyle stossichii* Braun, 1899, with an account of *Calicotyle palombi* sp. nov. *Parasitology 50*, 21–30.

Evans, G. C. 1976. A sack of uncut diamonds: The study of ecosystems and the future resources of mankind. *J. Ecol. 64*, 1–39.

Evans, J. W. 1962. Evolution of the Homoptera. In *The evolution of living organisms*. (Ed. by G. W. Leeper) pp. 250–259. Melbourne Univ. Press, Melbourne.

Feeny, P. 1976. Plant apparency and chemical defense. In *Biochemical interaction between plants and insects*. (Ed. by J. W. Wallace and R. L. Mansell), pp. 1–40. Rec. Adv. Phytochem. 10. Plenum, New York.

Fenner, F., and F. N. Ratcliffe. 1965. *Myxomatosis*. Cambridge Univ. Press, London.

Fenner, F., and D. O. White. 1976. *Medical virology*. 2nd ed. Academic Press, New York.

Fischer, M. 1962. Beitrag zur Kenntnis der Wirte von *Opius*-Arten. *Entomophaga 7*, 79–90.

Flessa, K. W., K. V. Powers, and J. L. Cisne. 1975. Specialization and evolutionary longevity in the Arthropoda. *Paleobiology 1*, 71–81.

Flor, H. H. 1956. The complementary genic systems in flax and flax rust. *Adv. Genet. 8*, 29–54.

Flor, H. H. 1971. Current status of the gene-for-gene concept. *Annu. Rev. Phytopathol. 9*, 272–296.

Ford, J. 1970. The geographical distribution of *Glossina*. In *The African trypanosomiases*. (Ed. by H. W. Mulligan), pp. 274–297. Wiley-Interscience, New York.

Ford, L. T. 1949. *A guide to the smaller British Lepidoptera*. South London Entomol. Natur. Hist. Soc., London.

Forrester, D. J. 1971. Bighorn sheep lungworm-pneumonia complex. In *Parasitic diseases of wild mammals*. (Ed. by J. W. Davis and R. C. Anderson), pp. 158–173. Iowa State Univ. Press, Ames, Iowa.

189

Foster, A. O. 1936. A quantitative study of the nematodes from a selected group of equines in Panama. *J. Parasitol. 22*, 479–510.

Foster, M. S. 1969. Synchronized life cycles in the orange-crowned warbler and its mallophagan parasites. *Ecology 50*, 315–323.

Frandsen, F. 1975. Host-parasite relationship of *Bulinus forskalii* (Ehrenberg) and *Schistosoma intercalatum* Fisher 1934, from Cameroun. *J. Helminth. 49*, 73–84.

Frankland, H.M.T. 1959. The incidence and distribution in Britain of the trematodes of *Talpa europaea*. *Parasitology 49*, 132–142.

Franklin, M. T. 1971. Taxonomy of Heteroderidae. In *Plant parasitic nematodes*. (Ed. by B. M. Zuckerman, W. F. Mai, and R. A. Rohde), 1:139–162. Academic Press, New York.

Freeland, W. J. 1976. Pathogens and the evolution of primate sociality. *Biotropica 8*, 12–24.

Frings, H., E. Goldberg, and J. C. Arentzen. 1948. Antibacterial action of the blood of the large milkweed bug. *Science 108*, 689–690.

Galun, R. 1975. The role of host blood in the feeding behavior of ectoparasites. In *Dynamic aspects of host-parasite relationships*. (Ed. by A. Zuckerman), 2:132–162. Keter, Jerusalem.

Garrett, S. D. 1970. *Pathogenic root-infecting fungi*. Cambridge Univ. Press, London.

Ghiselin, M. T. 1974. *The economy of nature and the evolution of sex*. Univ. California Press, Berkeley, Calif.

Gibson, C.W.D. 1976. The importance of foodplants for the distribution and abundance of some Stenodemini (Heteroptera: Miridae) of limestone grassland, *Oecologia 25*, 55–76.

Gilbert, L. E. 1975. Ecological consequences of a coevolved mutualism between butterflies and plants. In *Coevolution of animals and plants* (Ed. by L. E. Gilbert and P. H. Raven), pp. 210–240. Univ. Texas Press, Austin, Tex.

Gilbert, L. E., and M. C. Singer. 1975. Butterfly ecology. *Annu. Rev. Ecol. Syst. 6*, 365–397.

Gill, D. E. 1978. The metapopulation ecology of the red-spotted newt, *Notophthalmus viridescens* (Rafinesque). *Ecol. Monogr. 48*, 145–166.

Glazier, R., H. D. Crofton, and J. H. Whitlock. 1967. Differential hatching of eggs from morph variants of *Haemonchus contortus cayugensis* (Nematoda, Trichostrongylidae). *Cornell Vet. 57*, 194–200.

Glesener, R. R., and D. Tilman. 1978. Sexuality and the components of environmental uncertainty: Clues from geographic parthenogenesis in terrestrial animals. *Amer. Natur. 112*, 659–673.

Goodpasture, C. 1975. Comparative courtship behavior and karyology in *Monodontomerus* (Hymenoptera: Torymidae). *Ann. Entomol. Soc. Amer. 68*, 391–397.

Goodpasture, C., and E. E. Grissell. 1975. A karyological study of nine species of *Torymus* (Hymenoptera: Torymidae). *Can. J. Genet. Cytol. 17*, 413–422.

Gordh, G. 1975. Some evolutionary trends in the Chalcidoidea (Hymenoptera) with particular reference to host preference. *J. New York Entomol. Soc. 83*, 279–280.

Gould, S. J. 1974. Darwin's dilemma. *Natur. Hist. 83* (6), 16–22.

Gould, S. J. 1975. Darwin's "Big book." *Science 188*, 824–826.

Gould, S. J. 1977. *Ontogeny and phylogeny*. Belknap Press of Harvard Univ. Press, Cambridge, Mass.

191

Govier, R. N. 1966. *The inter-relationships of the hemiparasites and their hosts, with special reference to Odontites verna (Bell.) Dum.* Ph.D. Dissertation, Univ. of Wales.

Graham, K. 1967. Fungal-insect mutualism in trees and timber. *Annu. Rev. Entomol. 12,* 105–126.

Graham, M.W.R. de V. 1969. The Pteromalidae of North-Western Europe (Hymenoptera: Chalcidoidea). *Bull. Brit. Mus. (Natur. Hist.) Entomol. Suppl. 16,* 1–908.

Grant, V. 1966. The selective origin of incompatibility barriers in the plant genus *Gilia. Amer. Natur. 100,* 99–118.

Green, G. J. 1964. A color mutation, its inheritance, and the inheritance of pathogenicity in *Puccinia graminis* Pers. *Can. J. Bot. 42,* 1653–1664.

Green, G. J. 1966. Selfing studies with races 10 and 11 of wheat stem rust. *Can. J. Bot. 44,* 1255–1260.

Green, G. J. 1971. Hybridization between *Puccinia graminis tritici* and *Puccinia graminis secalis* and its evolutionary implications. *Can. J. Bot. 49,* 2089–2095.

Griffiths, G.C.D. 1964–1968. The Alysiinae (Hym. Braconidae) parasites of the Agromyzidae (Diptera). I–VI. *Beitr. Entomol. 14,* 823–914; *16,* 551–605; *16,* 775–951; *17,* 653–696; *18,* 5–62; *18,* 63–152.

Grubb, P. J. 1977. The maintenance of species-richness in plant communities: The importance of the regeneration niche. *Biol. Rev. 52,* 107–145.

Gurney, W.S.C., and R. M. Nisbet. 1978. Single-species population fluctuations in patchy environments. *Amer. Natur. 112,* 1075–1090.

Hair, J. D. 1975. *The structure of the intestinal helminth communities of lesser scaup (Aythya affinis).* Ph.D. Thesis, Univ. of Alberta.

Hair, J. D., and J. C. Holmes. 1975. The usefulness of measures of diversity niche width and niche overlap in the analysis of helminth communities of waterfowl. *Acta Parasit. Polonica 23*, 253–269.

Hairston, N. G. 1965. An analysis of age-prevalence data by catalytic models. *Bull. Wld. Hlth. Org. 33*, 163–175.

Halvörsen, O. 1976. Negative interaction amongst parasites. In *Ecological aspects of parasitology* (Ed. by C. R. Kennedy), pp. 99–114. North-Holland, Amsterdam.

Hamilton, W. D. 1978. Evolution and diversity under bark. *Symp. Roy. Entomol. Soc. London 9*, 154–175.

Harley, J. L. 1969. *The biology of mycorrhiza.* Leonard Hill, London.

Harper, J. L. 1977. *Population biology of plants.* Academic Press, London.

Harrison, L. 1914. The Mallophaga as a possible clue to bird phylogeny. *Austr. Zool. 1*, 7–11.

Hassell, M. P. 1976. Arthropod predator-prey systems. In *Theoretical ecology: Principles and applications* (Ed. by R. M. May), pp. 71–93. W. B. Saunders, Philadelphia.

Hassell, M. P. 1978. *The dynamics of arthropod predator-prey systems.* Princeton Univ. Press, Princeton, N.J.

Hastings, A. 1977. Spatial heterogeneity and the stability of predator-prey systems. *Theoret. Pop. Biol. 12*, 37–48.

Heatwole, H., D. M. Davis, and A. M. Wenner. 1964. Detection of mates and hosts by parasitic insects of the genus *Megarhyssa* (Hymenoptera: Ichneumonidae). *Amer. Midl. Natur. 71*, 374–381.

Hegnauer, R. 1971. Chemical patterns and relationships of Umbelliferae. In *The biology and chemistry of the Umbelliferae* (Ed. by V. H. Heywood), pp. 267–277. Academic Press, London.

Heinrich, G. H. 1960–1962. Synopsis of Nearctic Ichneumoninae Stenopneusticae with particular reference to the Northeastern Region (Hymenoptera) Parts I-VII. *Can. Entomol. Suppl. 15, 18, 21, 23, 26, 27, 29.*

Helle, W., and A. H. Pieterse. 1965. Genetic affinities between adjacent populations of spider mites (*Tetranychus urticae* Koch). *Entomol. Exp. Appl. 8,* 305–308.

Hering, E. M. 1957. *Bestimmungstabellen der Blattminen von Europa einschliesslich des Mittelmeerbeckens und der Kanarischen Inseln.* vols. 1–3. Junk, The Hague, Netherlands.

Hershkovitz, P. 1962. Evolution of neotropical cricetine rodents (Muridae) with special reference to the phyllotine group. *Fieldiana: Zoology 46,* 1–524.

Heywood, V. H. 1971. Systematic survey of Old World Umbelliferae. In *The biology and chemistry of the Umbelliferae* (Ed. by V. H. Heywood), pp. 31–41. Academic Press, London.

Hibler, C. P., R. E. Lange, and C. J. Metzger. 1972. Transplacental transmission of *Protostrongylus* spp. in bighorn sheep. *J. Wildl. Dis. 8,* 389.

Hirst, J. M. 1958. New methods for studying plant disease epidemics. *Outlook Agric. 2,* 16–26.

Holmes, J. C. 1961. Effects of concurrent infections on *Hymenolepis diminuta* (Cestoda) and *Moniliformis dubius* (Acanthocephala). I. General effects and comparison with crowding. *J. Parasitol. 47,* 209–216.

Holmes, J. C. 1962a. Effects of concurrent infections on *Hymenolepis diminuta* (Cestoda) and *Moniliformis dubius* (Acanthocephala). II. Effects on growth. *J. Parasitol. 48,* 87–96.

Holmes, J. C. 1962b. Effects of concurrent infections on *Hymenolepis diminuta* (Cestoda) and *Moniliformis dubius* (Acanthocephala). III. Effects in hamsters. *J. Parasitol. 48,* 97–100.

Holmes, J. C. 1971. Habitat segregation in sanguinicolid blood flukes (Digenea) of scorpaenid rockfishes (Perciformes) on the Pacific coast of North America. *J. Fish. Res. Bd. Can. 28*, 903–909.

Holmes, J. C. 1973. Site selection by parasitic helminths: interspecific interactions, site segregation, and their importance to the development of helminth communities. *Can. J. Zool. 51*, 333–347.

Holmes, J. C. 1976. Host selection and its consequences. In *Ecological aspects of parasitology* (Ed. by C. R. Kennedy), pp. 21–39. North-Holland, Amsterdam.

Holmes, J. C., and W. M. Bethel. 1972. Modification of intermediate host behaviour by parasites. In *Behavioural aspects of parasite transmission*. (Ed. by E. U. Canning and C. A. Wright), pp. 123–149. *Zool. J. Linn. Soc. 51*, Suppl. 1.

Holmes, J. C., R. P. Hobbs, and T. S. Leong. 1977. Populations in perspective: Community organization and regulation of parasite populations. In *Regulation of parasite populations* (Ed. by G. W. Esch), pp. 209–245. Academic Press, New York.

Hoogstraal, H. 1967. Ticks in relation to human diseases caused by *Rickettsia* species. *Annu. Rev. Entomol. 12*, 377–420.

Hopkins, G.H.E. 1942. The Mallophaga as an aid to the classification of birds. *Ibis 14*, 94–106.

Hopkins, G.H.E. 1949. The host-associations of the lice of mammals. *Proc. Zool. Soc. London 119*, 387–604.

Hopkins, G.H.E., and T. Clay. 1952. *A check list of the genera and species of Mallophaga*. Brit. Mus. (Natur. Hist.), London.

Hopkins, G.H.E., and M. Rothschild. 1962, 1966. *An illustrated catalogue of the Rothschild collection of fleas (Siphonaptera) in the British Museum (Natural His-*

tory). *Hystrichopsyllidae*, vol. 3. *Hystrichopsyllidae*, vol. 4. Trustees Brit. Mus., London.

Horn, H. S., and R. H. MacArthur. 1972. Competition among fugitive species in a harlequin environment. *Ecology 53*, 749–752.

Hovore, F. T., and E. F. Giesbert. 1976. Notes on the ecology and distribution of western Cerambycidae (Coleoptera). *Coleopterist's Bull. 30*, 349–360.

Huettel, M. D., and G. L. Bush. 1972. The genetics of host selection and its bearing on sympatric speciation in *Procecidochares* (Diptera: Tephritidae). *Entomol. Exp. Appl. 15*, 465–480.

Huffaker, C. B. 1958. Experimental studies on predation: dispersion factors and predator-prey oscillations. *Hilgardia 27*, 343–383.

Huffaker, C. B. 1964. Fundamentals of biological weed control. In *Biological control of insect pests and weeds*. (Ed. by P. DeBach), pp. 631–649. Reinhold, New York.

Hutchinson, G. E. 1953. The concept of pattern in ecology. *Proc. Acad. Nat. Sci. Philadelphia 105*, 1–12.

Hutchinson, G. E. 1957. Concluding remarks. *Cold Spring Harbor Symp. Quant. Biol. 22*, 415–427.

Hutchinson, G. E. 1959. Homage to Santa Rosalia or why are there so many kinds of animals? *Amer. Natur. 93*, 145–159.

Hutchinson, G. E. 1965. *The ecological theater and the evolutionary play*. Yale Univ. Press, New Haven, Conn.

Huxley, J. S. 1942. *Evolution, the modern synthesis*. Allen and Unwin, London.

Huxley, J. S. 1953. *Evolution in action*. Harper, New York.

Istock, C. A. 1967. Transient competitive displacement in natural populations of whirligig beetles. *Ecology 48*, 929–937.

Iwata, K. 1960. The comparative anatomy of the ovary in Hymenoptera. Part V., Ichneumonidae. *Acta Hymenopterologica 1*, 115–169.

Jaenike, J. 1978. An hypothesis to account for the maintenance of sex within populations. *Evol. Theory 3*, 191–194.

Jaenike, J., and R. K. Selander. 1979. Evolution and ecology of parthenogenesis in earthworms. *Amer. Zool.* 19:729–737.

Janzen, D. H. 1968. Host plants as islands in evolutionary and contemporary time. *Amer. Natur. 102*, 592–595.

Janzen, D. H. 1970. Herbivores and the number of tree species in tropical forests. *Amer. Natur. 104*, 501–528.

Janzen, D. H. 1973a. Host plants as islands. II. Competition in evolutionary and contemporary time. *Amer. Natur. 107*, 786–790.

Janzen, D. H. 1973b. Comments on host-specificity of tropical herbivores and its relevance to species richness. In *Taxonomy and ecology*. (Ed. by V. H. Heywood), pp. 201–211. Academic Press, New York.

Janzen, D. H. 1975. Interactions of seeds and their insect predators/parasitoids in a tropical deciduous forest. In *Evolutionary strategies of parasitic insects and mites*. (Ed. by P. W. Price), pp. 154–186. Plenum, New York.

Janzen, D. H., and C. M. Pond. 1975. A comparison, by sweep sampling, of the arthropod fauna of secondary vegetation in Michigan, England and Costa Rica. *Trans. Roy. Entomol. Soc. London. 127*, 33–50.

Jenkins, D., A. Watson, and G. R. Miller. 1963. Population studies on red grouse, *Lagopus lagopus scoticus* (Lath.) in north-east Scotland. *J. Anim. Ecol. 32*, 317–376.

Jenkins, D., A. Watson, and G. R. Miller. 1964. Predation and red grouse populations. *J. Appl. Ecol. 1*, 183–195.

197

Jennings, J. B., and P. Calow. 1975. The relationship between high fecundity and the evolution of entoparasitism. *Oecologia 21*, 109–115.

John, B. 1976. *Population cytogenetics*. Arnold, London.

Jones, A. W. 1967. *Introduction to parasitology*. Addison-Wesley, Reading, Mass.

Jordan, K. 1942. On *Parapsyllus* and some closely related genera of Siphonaptera. *Rev. Esp. Entomol. 18*, 7–29.

Kagan, I. G., and S. E. Maddison. 1975. Research studies on the immunology of schistosomiasis. In *Dynamic aspects of host-parasite relationships*. (Ed. by A. Zuckerman), 2: 163–176. Keter, Jerusalem.

Kao, K. N., and D. R. Knott. 1969. The inheritance of pathogenicity in races 111 and 29 of wheat stem rust. *Can. J. Genet. Cytol. 11*, 266–274.

Kellogg, V. L. 1913. Distribution and species-forming of ecto-parasites. *Amer. Natur. 47*, 129–158.

Kennedy, C. R. 1974. A checklist of British and Irish freshwater fish parasites with notes on their distribution. *J. Fish. Biol. 6*, 613–644.

Kennedy, C. R. 1975a. *Ecological animal parasitology*. Blackwell Sci. Pub., Oxford.

Kennedy, C. R. 1975b. The natural history of Slapton Ley Nature Reserve. VIII. The parasites of fish, with special reference to their use as a source of information about the aquatic community. *Field Stud. 4*, 177–189.

Kennedy, C. R. 1976. Reproduction and dispersal. In *Ecological aspects of parasitology*. (Ed. by C. R. Kennedy), pp. 143–160. North-Holland, Amsterdam.

Kennedy, C. R. 1977. The regulation of fish parasite populations. In *Regulation of parasite populations*. (Ed. by G. W. Esch), pp. 63–109. Academic Press, New York.

Kennedy, C. R., and A. Rumpus. 1977. Long-term changes in the size of the *Pomphorhynchus laevis* (Acanthocephala) population in the River Avon. *J. Fish. Biol.* *10*, 35–42.

Kethley, J. B., and D. E. Johnston. 1975. Resource tracking patterns in bird and mammal ectoparasites. *Misc. Publ. Entomol. Soc. Amer. 9*, 231–236.

Kinsey, A. C. 1937. An evolutionary analysis of insular and continental species. *Proc. Nat. Acad. Sci. U.S.A. 23*, 5–11.

Kloet, G. S., and W. D. Hincks. 1945. *A check list of British insects*. Kloet and Hincks, Stockport.

Klopfer, P. H. 1959. Environmental determinants of faunal diversity. *Amer. Natur. 93*, 337–342.

Klopfer, P. H., and R. H. MacArthur. 1960. Niche size and faunal diversity. *Amer. Natur. 94*, 293–300.

Koford, C. B. 1958. Prairie dogs, whitefaces, and blue grama. *Wildl. Monogr. 3*, 1–78.

Kostrowicki, A. S. 1969. *Geography of the Palearctic Papilionoidea (Lepidoptera)*. Panstwowe Wydawnictwo Naukowe, Krakow.

Krebs, C. J. 1972. *Ecology. The experimental analysis of distribution and abundance*. Harper and Row, New York.

Kuijt, J. 1969. *The biology of parasitic flowering plants*. Univ. California Press, Berkeley, Calif.

Large, E. C. 1940. *The advance of the fungi*. Cape, London. Reprint 1962. Dover, New York.

Lawton, J. H. 1976. The structure of the arthropod community on bracken. *Bot. J. Linnean Soc. 73*, 187–216.

Lawton, J. H. 1978. Host-plant influences on insect diversity: the effects of space and time. *Symp. Roy. Entomol. Soc. London. 9*, 105–125.

Lawton, J. H., and P. W. Price. 1979. Species richness of parasites on hosts: Agromyzid flies on the British Umbelliferae. *J. Anim. Ecol. 48*, 619–637.

Lawton, J. H., and D. Schröder. 1977. Effects of plant type, size of geographical range and taxonomic isolation on number of insect species associated with British plants. *Nature 265*, 137–140.

Lees, A. D. 1955. *The physiology of diapause in arthropods.* Cambridge Univ. Press, London.

LeJambre, L. F., and L. H. Ratcliffe. 1971. Seasonal change in a balanced polymorphism in *Haemonchus contortus* populations. *Parasitology 62*, 151–155.

LeJambre, L. F., L. H. Ratcliffe, J. H. Whitlock, and H. D. Crofton. 1970. Polymorphism and egg-size in the sheep nematode, *Haemonchus contortus. Evolution 24*, 625–631.

LeJambre, L. F., L. H. Ratcliffe, L. S. Uhazy, and J. H. Whitlock. 1972. Evidence that the seasonal alternation of polymorphs apparent in linguiform *H. contortus cayugensis* also occurs in the other phenotypes. *Int. J. Parasitol. 2*, 171–173.

LeJambre, L. F., and J. H. Whitlock. 1968. Seasonal fluctuation in linguiform morphs of *Haemonchus contortus cayugensis. J. Parasitol. 54*, 827–830.

LeJambre, L. F., and J. H. Whitlock. 1976. Changes in the hatch rate of *Haemonchus contortus* eggs between geographic regions. *Parasitology, 73*, 223–238.

Lejeune, R. R., and V. Hildahl. 1954. A survey of parasites of the larch sawfly [*Pristiphora erichsonii* (Hartig)] in Manitoba and Saskatchewan. *Can. Entomol. 86*, 337–345.

Levin, D. A. 1975. Pest pressure and recombination systems in plants. *Amer. Natur. 109*, 437–451.

Levin, S. A. 1976. Population dynamic models in heterogeneous environments. *Annu. Rev. Ecol. Syst. 7*, 287–310.

Levins, R. 1962. Theory of fitness in a heterogeneous environment. I. The fitness set and adaptive function. *Amer. Natur. 96*, 361–373.

Levins, R., and D. Culver. 1971. Regional coexistence of species and competition between rare species. *Proc. Nat. Acad. Sci. U.S.A. 68*, 1246–1248.

Lewis, T. 1973. *Thrips, their biology, ecology and economic importance*. Academic Press, London.

Limbaugh, C. 1961. Cleaning symbiosis. *Sci. Amer. 205* (2), 42–49.

Lin, N., and C. D. Michener. 1972. Evolution of sociality in insects. *Quart. Rev. Biol. 47*, 131–159.

Lindquist, E. E. 1969. Review of holarctic tarsonemid mites (Acarina: Prostigmata) parasitizing eggs of ipine bark beetles. *Mem. Entomol. Soc. Can. 60*, 1–111.

Llewellyn, J. 1956. The host-specificity, micro-ecology, adhesive attitudes, and comparative morphology of some trematode gill parasites. *J. Mar. Biol. Ass. U. K. 35*, 113–127.

Lloyd, M., and H. S. Dybas. 1966. The periodical cicada problem. I. Population ecology. *Evolution 20*, 133–149.

Loegering, W. Q., and H. R. Powers, Jr. 1962. Inheritance of pathogenicity in a cross of physiological races 111 and 36 of *Puccinia graminis* f. sp. *tritici. Phytopathology 52*, 547–554.

Luig, N. H., and I. A. Watson. 1961. A study of inheritance of pathogenicity in *Puccinia graminis* var. *tritici. Proc. Linn. Soc. N. S. W. 86*, 217–229.

MacArthur, R. H. 1957. On the relative abundance of bird species. *Proc. Nat. Acad. Sci. U.S.A. 43*, 293–295.

MacArthur, R. H. 1968. The theory of the niche. In *Population biology and evolution*. (Ed. by R. C. Lewontin), pp. 159–176. Syracuse Univ. Press, Syracuse, N.Y.

MacArthur, R. H. 1972. *Geographical ecology*. Harper & Row, N.Y.

MacArthur, R. H., and R. Levins. 1964. Competition, habitat selection, and character displacement in a patchy environment. *Proc. Nat. Acad. Sci. U.S.A. 51*, 1207–1210.

MacArthur, R. H., and E. O. Wilson. 1967. *The theory of island biogeography*. Princeton Univ. Press, Princeton, N.J.

MacInnis, A. J. 1976. How parasites find hosts: Some thoughts on the inception of host-parasite integration. In *Ecological aspects of parasitology*. (Ed. by C. R. Kennedy), pp. 3–39. North-Holland, Amsterdam.

MacKenzie, K., and D. I. Gibson. 1970. Ecological studies of some parasites of plaice *Pleuronectes platessa* L. and flounder *Platichthys flesus* (L.). In *Aspects of fish parasitology*. (Ed. by A.E.R. Taylor and R. Muller), pp. 1–42. Symp. Brit. Soc. Parasitol. 8.

Malavasi, A., A. B. daCunha, J. S. Morgante, and J. Marques. 1976. Relationships between the gregarine *Schneideria schneiderae* and its host *Trichosia pubescens* (Diptera, Sciaridae). *J. Invertebr. Pathol. 28*, 363–371.

Manter, H. W. 1966. Parasites of fishes as biological indicators of recent and ancient conditions. In *Host-parasite relationships*. (Ed. by J. E. McCauley), pp. 59–71. Oregon State Univ. Press, Corvallis, Oreg.

Marchalonis, J. J. 1977. *Immunity in evolution*. Harvard Univ. Press, Cambridge, Mass.

Marks, G. C., F. Y. Kassaby, and P. C. Fagg. 1975. Variation in population levels of *Phytophthora cinnamomi* in *Eucalyptus* forest soils of eastern Victoria. *Aust. J. Bot. 23*, 435–449.

Marshall, A. G. 1976. Host-specificity amongst arthropods ectoparasitic upon mammals and birds in the New Hebrides. *Ecol. Entomol. 1*, 189–199.

Martens, J. W., R.I.H. McKenzie, and G. J. Green. 1970. Gene-for-gene relationships in the *Avena: Puccinia graminis* host-parasite system in Canada. *Can. J. Bot.* *48*, 969–975.

Martin, A. C., H. S. Zim, and A. L. Nelson. 1951. *American wildlife and plants.* McGraw-Hill, New York.

Martin, D. R. 1969. Lecithodendriid trematodes from the bat *Peropteryx kappleri* in Columbia, including discussions of allometric growth and significance of ecological isolation. *Proc. Helminthol. Soc. Wash. 36,* 250–260.

May, R. M. 1973. *Stability and complexity in model ecosystems.* Princeton Univ. Press, Princeton, N.J.

May, R. M. 1977a. Dynamical aspects of host-parasite associations: Crofton's model revisited. *Parasitology 75,* 259–276.

May, R. M. 1977b. Togetherness among schistosomes: Its effects on the dynamics of the infection. *Math. BioSci. 35,* 301–343.

May, R. M. 1978a. Host-parasitoid systems in patchy environments: A phenomenological model. *J. Anim. Ecol. 47,* 833–843.

May, R. M. 1978b. The dynamics and diversity of insect faunas. *Symp. Roy. Entomol. Soc. London 9,* 188–204.

May, R. M., and R. M. Anderson. 1978. Regulation and stability of host-parasite population interactions. II. Destabilizing processes. *J. Anim. Ecol. 47,* 249–267.

Maynard Smith, J. 1966. Sympatric speciation. *Amer. Natur. 100,* 637–650.

Maynard Smith, J. 1978. *The evolution of sex.* Cambridge Univ. Press, Cambridge.

Mayr, E. 1963. *Animal species and evolution.* Belknap Press of Harvard Univ. Press, Cambridge, Mass.

Mayr, E. 1970. *Populations, species and evolution.* Belknap Press of Harvard Univ. Press, Cambridge, Mass.

Mayr, E. 1974. The challenge of diversity. *Taxon 23,* 3–9.

Mayr, E. 1976. *Evolution and the diversity of life. Selected essays.* Belknap Press of Harvard Univ. Press, Cambridge, Mass.

Mayr, E., E. G. Linsley, and R. L. Usinger. 1953. *Methods and principles of systematic zoology.* McGraw-Hill, New York.

Mayse, M. A., and P. W. Price. 1978. Seasonal development of soybean arthropod communities in east central Illinois. *Agro-ecosystems 4,* 387–405.

McClure, M. S., and P. W. Price. 1976. Ecotope characteristics of coexisting *Erythroneura* leafhoppers (Homoptera: Cicadellidae) on sycamore. *Ecology 57,* 928–940.

McLeod, J. M. 1972. The Swaine jack pine sawfly, *Neodiprion swainei,* life system: Evaluating the long-term effects of insecticide applications in Quebec. *Environ. Entomol. 1,* 371–381.

McNeill, W. H. 1977. *Plagues and peoples.* Anchor Press, Garden City, N.Y.

Melander, L. W., and J. H. Craigie. 1927. Nature of resistance of *Berberris* spp. to *Puccinia graminis. Phytopathology 17,* 95–114.

Mellinger, M. V., and S. J. McNaughton. 1975. Structure and function of successional vascular plant communities in central New York. *Ecol. Monogr. 45,* 161–182.

Metcalf, M. M. 1929. Parasites and the aid they give in problems of taxonomy, geographical distribution, and paleogeography. *Smithsonian Misc. Coll. 81* (8), 1–36.

Michener, C. D. 1958. The evolution of social behavior in bees. *Proc. 10th Int. Congr. Entomol., Montreal 2,* 441–447.

Mitchell, R. 1970. An analysis of dispersal in mites. *Amer. Natur. 104*, 425–431.

Mitter, C., and D. Futuyma. 1977. Parthenogenesis in the fall cankerworm, *Alsophila pometaria* (Lepidoptera, Geometridae). *Entomol. Exp. Appl. 21*, 192–198.

Mitter, C., D. J. Futuyma, J. C. Schneider, and J. D. Hare. 1979. Genetic variation and host plant relations in a parthenogenetic moth. *Evolution 33*, 777–790.

Moericke, V., R. J. Prokopy, S. Berlocher, and G. L. Bush. 1975. Visual stimuli eliciting attraction of *Rhagoletis pomonella* (Diptera: Tephritidae) flies to trees. *Ent. Exp. Appl. 18*, 497–507.

Mook, J. H. 1967. Habitat selection by *Lipara lucens* Mg. (Diptera, Chloropidae) and its survival value. *Arch. Néerl. Zool. 17*, 469–549.

Mook, J. H. 1971. Influence of environment on some insects attacking common reed *(Phragmites communis* Trin.). *Hidrobiologia 12*, 305–312.

Moore, W. C. 1959. *British parasitic fungi.* Cambridge Univ. Press, London.

Morris, J. R. 1970. *An ecological study of the basommatophoran snail Helisoma trivolvis in central Alberta.* Ph.D. Thesis, Univ. of Alberta.

Morris, M. G. 1974. Oak as a habitat for insect life. In *The British oak.* (Ed. by M. G. Morris and F. H. Perring), pp. 274–297. Classey, Faringdon, Berks.

Morris, R. F. 1969. Approaches to the study of population dynamics. In *Forest insect population dynamics.* (Ed. by W. E. Waters), pp. 9–28. U.S.D.A. Forest Serv. Res. Pap. NE-125, Upper Darby, Penn.

Morris, R. F. 1976. Relation of parasite attack to the colonial habit of *Hyphantria cunea. Can. Entomol. 108*, 833–836.

Morrow, P. A. 1977a. Host specificity of insects in a community of three co-dominant *Eucalyptus* species. *Aust. J. Ecol. 2*, 89–106.

Morrow, P. A. 1977b. The significance of phytophagous insects in the *Eucalyptus* forests of Australia. In *The role of arthropods in forest ecosystems.* (Ed. by W. J. Mattson), pp. 19–29. Springer, New York.

Morrow, P. A., and V. C. LaMarche, Jr. 1978. Tree ring evidence for chronic insect suppression of productivity in subalpine *Eucalyptus. Science 201,* 1244–1246.

Mulligan, H. W. (ed.). 1970. *The African trypanosomiases.* Wiley-Interscience, New York.

Murdoch, W. W., F. C. Evans, and C. H. Peterson. 1972. Diversity and pattern in plants and insects. *Ecology 53,* 819–829.

Murray, J. S. 1974. The fungal pathogens of oak. In *The British oak.* (Ed. by M. G. Morris and F. H. Perring), pp. 235–249. Classey, Faringdon, Berks.

Naumov, N. P. 1972. *The ecology of animals.* (Trans. F. K. Plous, Jr.). Univ. Illinois Press, Urbana, Ill.

Nei, M. 1975. *Molecular population genetics and evolution.* Elsevier, New York.

Nelson, B. C., and M. D. Murray. 1971. The distribution of Mallophaga on the domestic pigeon *(Columba livia). Int. J. Parasitol. 1,* 21–29.

Nelson, B. C., and C. R. Smith. 1976. Ecological effects of a plague epizootic on the activities of rodents inhabiting caves at Lava Beds National Monument, California. *J. Med. Entomol. 13,* 51–61.

Newell, N. D. 1949. Phyletic size increase, an important trend illustrated by fossil invertebrates. *Evolution 3,* 103–124.

Newell, N. D. 1967. Revolutions in the history of life. *Geol. Soc. Amer. Spec. Paper. 89,* 63–91.

Noble, E. R., and G. A. Noble. 1976. *Parasitology. The biology of animal parasites.* 4th ed. Lea and Febiger, Philadelphia.

Nowakowski, J. T. 1962. Introduction to a systematic revision of the family Agromyzidae (Diptera) with some remarks on host-plant selection by these flies. *Ann. zool. Warszawa 20* (8), 67–183.

Odum, E. P. 1971. *Fundamentals of ecology.* 3rd ed. W. B. Saunders, Philadelphia.

Oliver, J. H., Jr. 1971. Parthenogenesis in mites and ticks (Arachnida: Acari). *Amer. Zool. 11,* 283–299.

Opler, P. A. 1974. Oaks as evolutionary islands for leaf-mining insects. *Amer. Sci. 62,* 67–73.

Otte, D., and A. Joern. 1976. On feeding patterns in desert grasshoppers and the evolution of specialized diets. *Proc. Acad. Nat. Sci. Phil. 128,* 89–126.

Owen, D. F., and J. Owen. 1974. Species diversity in temperate and tropical Ichneumonidae. *Nature 249,* 583–584.

Owen, D. F., and R. G. Wiegert. 1976. Do consumers maximize plant fitness? *Oikos 27,* 488–492.

Pamilo, P., K. Vespäläinen, and R. Rosengren. 1975. Low allozymic variability in *Formica* ants. *Hereditas 80,* 293–296.

Paperna, I. 1964. Competitive exclusion of *Dactylogyrus extensus* by *Dactylogyrus vastator* (Trematoda: Monogenea) on the gills of reared carp. *J. Parasitol. 50,* 94–98.

Park, T. 1948. Experimental studies of interspecies competition. I. Competition between populations of the flour beetles, *Tribolium confusum* Duval and *Tribolium castaneum* Herbst. *Ecol. Monogr. 18,* 265–308.

Parker, E. D., Jr., R. K. Selander, R. O. Hudson, and L. J.

Lester. 1977. Genetic diversity in colonizing partheno-genetic cockroaches. *Evolution 31*, 836–842.

Pearsall, W. H. 1954. Biology and land-use in East Africa. *New Biol.* 17:9–26.

Perring, F. H., and S. M. Walters. 1962. *Atlas of the British flora*. Bot. Soc. Brit. Is. Nelson, London.

Petter, A. J. 1962. Redescription et analyse critique de quelque espèces d'Oxyures de la tortue grecque (*Testudo graeca* L.). Diversité des structures céphaliques (II). *Ann. Parasitol. Hum. Comp. 37*, 140–152.

Petter, A. J. 1966. Équilibre des espèces dans les populations de nématodes parasites du côlon des tortues terrestres. *Mém. Mus. Natl. Hist. Nat. Paris Ser. A. Zool. 39* (1), 1–252.

Phillips, P. A., and M. M. Barnes. 1975. Host race formation among sympatric apple, plum and walnut populations of the codling moth, *Laspeyresia pomonella*. *Ann. Entomol. Soc. Amer. 68*, 1053–1068.

Pianka, E. R. 1966. Latitudinal gradients in species diversity: A review of concepts. *Amer. Natur. 100*, 33–46.

Pianka, E. R. 1974. *Evolutionary ecology*. Harper & Row, New York.

Pickett, S.T.A. 1976. Succession: An evolutionary interpretation. *Amer. Natur. 110*, 107–119.

Pickett, S.T.A., and J. N. Thompson. 1978. Patch dynamics and the design of nature reserves. *Biol. Conserv. 13*, 27–37.

Pielou, E. C. 1969. *An introduction to mathematical ecology*. Wiley-Interscience, New York.

Pimentel, D., and A. C. Bellotti. 1976. Parasite-host population systems and genetic stability. *Amer. Natur. 110*, 877–888.

Pimm, S. L., and J. H. Lawton. 1977. Number of trophic levels in ecological communities. *Nature 268*, 329–331.

Poole, R. W. 1974. *An introduction to quantitative ecology.* McGraw-Hill, New York.

Powell, J. A., and R. A. Mackie. 1966. Biological interrelationships of moths and *Yucca whipplei. Univ. Calif. Publ. Entomol. 42,* 1–59.

Powell, J. R. 1978. The founder-flush speciation theory: An experimental approach. *Evolution 32,* 465–474.

Powell, J. R., and H. Wistrand. 1978. The effect of heterogeneous environments and a competitor on genetic variation in *Drosophila. Amer. Natur. 112,* 935–947.

Powell, N. T. 1971. Interaction of plant parasitic nematodes with other disease-causing agents. In *Plant parasitic nematodes.* (Ed. by B. M. Zuckerman, W. F. Mai, and R. A. Rohde), 2:119–136. Academic Press, New York.

Price, P. W. 1973a. The development of parasitoid communities. *Proc. Northeastern Forest Insect Work Conf. 5,* 29–41.

Price, P. W. 1973b. Parasitoid strategies and community organization. *Environ. Entomol. 2,* 623–626.

Price, P. W. 1974a. Strategies for egg production. *Evolution 28,* 76–84.

Price, P. W. 1974b. Energy allocation in ephemeral adult insects. *Ohio J. Sci. 74,* 380–387.

Price, P. W. (ed.). 1975a. *Evolutionary strategies of parasitic insects and mites.* Plenum, New York.

Price, P. W. 1975b. Reproductive strategies of parasitoids. In *Evolutionary strategies of parasitic insects and mites.* (Ed. by P. W. Price), pp. 87–111. Plenum, New York.

Price, P. W. 1976. Colonization of crops by arthropods: Non-equilibrium communities in soybean fields. *Environ. Entomol. 5,* 605–611.

209

Price, P. W. 1977. General concepts on the evolutionary biology of parasites. *Evolution 31*, 405–420.

Price, P. W. 1979. The extent of adaptive radiation in fleas (Siphonaptera). In *Proc. Int. Conf. Fleas*. (Ed. by R. Traub). In press. Balkema, Rotterdam.

Price, P. W., and M. F. Willson. 1976. Some consequences for a parasitic herbivore, the milkweed longhorn beetle, *Tetraopes tetrophthalmus*, of a host-plant shift from *Asclepias syriaca* to *A. verticillata*. *Oecologia 25*, 331–340.

Prokopy, R. J. 1968. Visual responses of apple maggot flies, *Rhagoletis pomonella* (Diptera: Tephritidae): Orchard Studies. *Entomol. Exp. Appl. 11*, 403–422.

Prop, N. 1960. Protection against birds and parasites in some species of tenthredinid larvae. *Arch. Néerl. Zool. 13*, 380–447.

Pyke, G. H., H. R. Pulliam, and E. L. Charnov. 1977. Optimal foraging: A selective review of theory and tests. *Quart. Rev. Biol. 52*, 137–154.

Quednau, F. W., and M. Guevrement. 1975. Observations on mating and oviposition behaviour of *Priopoda nigricollis* (Hymenoptera: Ichneumonidae), a parasite of the birch leaf-miner, *Fenusa pusilla* (Hymenoptera: Tenthredinidae). *Can. Entomol. 107*, 1119–1204.

Rathcke, B. J. 1976. Insect-plant patterns and relationships in the stem-boring guild. *Amer. Midl. Natur. 96*, 98–117.

Rathcke, B. J., and P. W. Price. 1976. Anomalous diversity of tropical ichneumonid parasitoids: A predation hypothesis. *Amer. Natur. 110*, 889–893.

Raup, D. M., S. J. Gould, T.J.M. Schopf, and D. S. Simberloff. 1973. Stochastic models of phylogeny and the evolution of diversity. *J. Geol. 81*, 525–542.

Rhoades, D. F., and R. G. Cates. 1976. Toward a general

theory of plant antiherbivore chemistry. In *Biochemical interaction between plants and insects*. (Ed. by J. W. Wallace and R. L. Mansell), pp. 168–213. Rec. Adv. Phytochem. 10. Plenum, London.

Richards, C. S. 1976. Genetics of the host-parasite relationship between *Biomphalaria glabrata* and *Schistosoma mansoni*. In *Genetic aspects of host-parasite relationships*. (Ed. by A.E.R. Taylor and R. Muller), pp. 45–54. Symp. Brit. Soc. Parasitol. 14.

Richards, P. W. 1952. *The tropical rain forest: An ecological study*. Cambridge Univ. Press, London.

Riom, J., and J. P. Fabre. 1977. Étude biologique et écologique de la cochenille du pin maritime, *Matsucoccus feytaudi* Ducasse, 1942 (Coccoidea, Margarodidae, Xylococcinae) dans le sud-est de la France. II. Régulation du cycle annuel comportements des stades mobiles. *Ann. Zool. Écol. Anim. 9*, 181–209.

Rishbeth, J. 1955. *Fusarium* wilt of bananas of Jamaica. I. Some observations on the epidemiology of the disease. *Ann. Bot. N. S. 19*, 293–328.

Rohde, K. 1976. Marine parasitology in Australia. *Search 7*, 477–482.

Rohde, K. 1977a. A non-competitive mechanism responsible for restricting niches. *Zool. Anz. 199*, 164–172.

Rohde, K. 1977b. Species diversity of monogenean gill parasites of fish on the Great Barrier Reef. *Proc. 3rd Int. Coral Reef Symp.*, 585–591. Univ. of Miami, Fla.

Rohde, K. 1977c. Habitat partitioning in Monogenea of marine fishes: *Heteromicrocotyla australiensis*, sp. nov. and *Heteromicrocotyloides mirabilis*, gen. and sp. nov. (Heteromicrocotylidae) on the gills of *Carangoides emburyi* (Carangidae) on the Great Barrier Reef, Australia. *Z. Parasitenkd. 53*, 171–182.

Rohde, K. 1978. Latitudinal differences in host-specificity

of marine Monogenea and Digenea. *Marine Biol. 47*, 125–134.

Root, R. B. 1975. Some consequences of ecosystem texture. In *Ecosystem analysis and prediction*. (Ed. by S. A. Levin), pp. 83–97. Proc. Soc. Ind. Appl. Math., Philadelphia.

Root, R. B., and S. J. Chaplin. 1976. The life-styles of tropical milkweed bugs, *Oncopeltus* (Hemiptera: Lygaeidae) utilizing the same hosts. *Ecology 57*, 132–140.

Rosenfield, P. L., R. A. Smith, and M. G. Wolman. 1977. Development and verification of a schistosomiasis transmission model. *Amer. J. Trop. Med. Hyg. 26*, 505–516.

Ross, H. H. 1962. *A synthesis of evolutionary theory*. Prentice-Hall, Englewood Cliffs, N.J.

Roth, L. M. 1967. Sexual isolation in parthenogenetic *Pycnoscelus surinamensis* and application of the name *Pycnoscelus indicus* to its bisexual relative (Dictyoptera: Blattaria: Blaberidae: Pycnoscelinae). *Ann. Entomol. Soc. Amer. 60*, 774–779.

Roth, L. M. 1974. Reproductive potential of bisexual *Pycnoscelus indicus* and clones of its parthenogenetic relative, *Pycnoscelus surinamensis. Ann. Entomol. Soc. Amer. 67*, 215–223.

Roth, L. M., and S. H. Cohen. 1968. Chromosomes of the *Pycnoscelus indicus* and *P. surinamensis* complex (Blattaria: Blaberidae: Pycnoscelinae). *Psyche 75*, 53-76.

Rothschild, M. 1965. The rabbit flea and hormones. *Endeavour 24*, 162–168.

Rothschild, M., and B. Ford. 1964. Maturation and egg-laying of the rabbit flea (*Spilopsyllus cuniculi* Dale) induced by the external application of hydrocortisone. *Nature 203*, 210–211.

Rothschild, M., and B. Ford. 1973. Factors influencing the breeding of the rabbit flea (*Spilopsyllus cuniculi*): A spring-time accelerator and a kairomone in nesting rabbit urine with notes on *Cediopsylla simplex*, another "hormone bound" species. *J. Zool. Lond. 170*, 87–137.

Royama, T. 1970. Factors governing the hunting behaviour and selection of food by the great tit (*Parus major* L.). *J. Anim. Ecol. 39*, 619–668.

Salt, G. 1941. The effects of hosts upon their insect parasites. *Biol. Rev. 16*, 239-264.

Salt, G. 1961. Competition among insect parasitoids. *Symp. Soc. Exp. Biol. 15*, 96–119.

Sankurathri, C. S., and J. C. Holmes. 1976a. Effects of thermal effluents on the population dynamics of *Physa gyrina* Say (Mollusca: Gastropoda) at Lake Wabamun, Alberta. *Can. J. Zool. 54*, 582–590.

Sankurathri, C. S., and J. C. Holmes. 1976b. Effects of thermal effluents on parasites and commensals of *Physa gyrina* Say (Mollusca: Gastropoda) and their interactions at Lake Wabamun, Alberta. *Can. J. Zool. 54*, 1742–1753.

Saville, D.B.O. 1975. Evolution and biogeography of Saxifragaceae with guidance from their rust parasites. *Ann. Missouri Bot. Gard. 62*, 354–361.

Schad, G. A. 1956. Studies on the genus *Kalicephalus* (Nematoda: Diaphanocephalidae). I. On the life histories of the North American species *K. parvus*, *K. agkistrodontis*, and *K. rectiphilus. Can. J. Zool. 34*, 425–452.

Schad, G. A. 1962a. Gause's hypothesis in relation to the oxyuroid populations of *Testudo graeca. J. Parasitol. 48* Sect. 2 (suppl.), 36–37.

Schad, G. A. 1962b. Studies on the genus *Kalicephalus* (Nematoda: Diaphanocephalidae). II. A taxonomic revision of the genus *Kalicephalus* Molin, 1861. *Can. J. Zool. 40*, 1035–1165.

Schad, G. A. 1963a. Niche diversification in a parasite species flock. *Nature 198*, 404–406.

Schad, G. A. 1963b. The ecology of co-occurring congeneric pinworms in tortoise, *Testudo graeca. Proc. 16th Int. Congr. Zool. 1*, 223–224.

Schad, G. A., R. E. Kuntz, and W. H. Wells. 1960. Nematode parasites from Turkish vertebrates. An annotated list. *Can. J. Zool. 38*, 949–963.

Schaller, G. B. 1972. *The Serengeti lion. A study of predator-prey relations.* Univ. of Chicago Press, Chicago.

Schnathorst, W. C., and J. E. DeVay. 1963. Common antigens in *Xanthomonas malvacearum* and *Gossypium hirsutum* and their possible relationship to host specificity and disease resistance. *Phytopathology 53*, 1142.

Schneider, J. C. 1980. The role of parthenogenesis and female aptery in microgeographic ecological variation in the fall cankerworm *Alsophila pometaria* Har. (Lepidoptera: Geometridae). *Ecology.* In Press.

Schoener, T. W. 1974. Resource partitioning in ecological communities. *Science 185*, 27–39.

Schopf, T.J.M., D. M. Raup, S. J. Gould, and D. S. Simberloff. 1975. Genomic versus morphologic rates of evolution: influence of morphologic complexity. *Paleobiology 1*, 63–70.

Scott, G. R. 1970. Rinderpest. In *Infectious diseases of wild mammals.* (Ed. by J. W. Davis, L. H. Karstad, and D. O. Trainer), pp. 20–35. Iowa State Univ. Press, Ames, Iowa.

Scriber, J. M. 1973. Latitudinal gradients in larval feeding specialization of the World Papilionidae (Lepidoptera). *Psyche 80*, 355–373.

Scudder, G.G.E. 1974. Species concepts and speciation. *Can. J. Zool. 52*, 1121–1134.

Selander, R. K. 1976. Genic variation in natural populations. In *Molecular evolution*. (Ed. by F. J. Ayala), pp. 21–45. Sinauer, Sunderland, Mass.

Shapiro, A. M. 1976. Beau Geste? *Amer. Natur. 110*, 900–902.

Sharp, M. A., D. R. Parks, and P. R. Ehrlich. 1974. Plant resources and butterfly habitat selection. *Ecology 55*, 870–875.

Shattock, R. C. 1977. The dynamics of plant diseases. In *Origins of pest, parasite, disease and weed problems*. (Ed. by J. M. Cherrett and G. R. Sagar), pp. 83–107. Blackwell Sci. Pub., Oxford.

Shields, O. 1967. Hilltopping. An ecological study of summit congregation behavior of butterflies on a southern California hill. *J. Res. Lepid. 6*, 69–178.

Simberloff, D. S., and E. O. Wilson. 1969. Experimental zoogeography of islands: the colonization of empty islands. *Ecology 50*, 278–296.

Simberloff, D. S., and E. O. Wilson. 1970. Experimental zoogeography of islands. A two-year record of colonization. *Ecology 51*, 934–937.

Simmonds, H. W. 1934. Biological control of noxious weeds, with special reference to the plants *Clidemia hirta* (the curse) and *Stachytarpheta jamaicensis* (blue rat tail). *Fiji Dept. Agric. J. 7*, 3–10.

Simpson, G. G. 1953. *The major features of evolution*. Columbia Univ. Press, New York.

Singer, M. C. 1971. Evolution of food-plant preference in the butterfly *Euphydryas editha*. *Evolution 25*, 383–389.

Skellam, J. G. 1951. Random dispersal in theoretical populations. *Biometrica 38*, 196–218.

Slansky, F., Jr., 1976. Phagism relationships among butterflies. *J. N. Y. Entomol. Soc. 84*, 91–105.

Slatkin, M. 1974. Competition and regional coexistence. *Ecology 55*, 128–134.

Smith, N. G. 1968. The advantage of being parasitized. *Nature 219*, 690–694.

Smithers, S. R., and M. J. Worms. 1976. Blood fluids—Helminths. In *Ecological aspects of parasitology*. (Ed. by C. R. Kennedy), pp. 349–369. North-Holland, Amsterdam.

Snyder, T. P. 1974. Lack of allozymic variability in three bee species. *Evolution 28*, 687–689.

Sogandares-Bernal, F. 1959. Digenetic trematodes of marine fishes from the Gulf of Panama and Bimini, British West Indies. *Tulane Stud. Zool. 7*, 69–117.

Sonneborn, T. M. 1957. Breeding systems, reproductive methods, and species problems in Protozoa. In *The species problem*. (Ed. by E. Mayr), pp. 155–324. Amer. Assoc. Adv. Sci. no. 50., Washington, D.C.

Southwood, T.R.E. 1961. The number of species of insect associated with various trees. *J. Anim. Ecol. 30*, 1–8.

Southwood, T.R.E. 1977. The stability of the trophic milieu, its influence on the evolution of behaviour and of responsiveness of trophic signals. *Colloq. Int. Cent. Natl. Rech. Sci. 265*, 471–493.

Southwood, T.R.E., and D. Leston. 1959. *Land and water bugs of the British Isles*. Warne, London.

Southwood, T.R.E., and G.G.E. Scudder. 1956. The immature stages of the Hemiptera-Heteroptera associated with the stinging nettle (*Urtica dioica* L.). *Entomol. Mon. Mag. 92*, 313–325.

Spencer, K. A. 1972. *Handbooks for the identification of British insects*. vol. 10. Part 5 (g) *Diptera, Agromyzidae*. Roy. Entomol. Soc. London, London.

Spieth, H. T. 1968. Evolutionary implications of sexual behavior in *Drosophila*. *Evol. Biol. 2*, 157–193.

Spinage, C. A. 1962. Rinderpest and faunal distribution patterns. *African Life 16*, 55.

Stebbins, G. L. 1958. Longevity, habitat and release of genetic variability in the higher plants. *Cold Spring Harbor Symp. Quant. Biol. 23*, 365–378.

Stebbins, G. L. 1964. The evolution of animal species. *Evolution 18*, 134–137.

Stebbins, G. L. 1974. *Flowering plants. Evolution above the species level.* Belknap Press of Harvard Univ. Press, Cambridge, Mass.

Steinhaus, E. A. 1946. *Insect microbiology.* Comstock, Ithaca, N.Y.

Stelfox, J. G. 1971. Bighorn sheep in the Canadian Rockies: A history 1800–1970. *Can. Field-Natur. 85*, 101–122.

Stern, K., and L. Roche. 1974. *Genetics of forest ecosystems.* Springer-Verlag. Berlin.

Stevenson-Hamilton, J. 1957. Tsetse fly and the rinderpest epidemic of 1896. *S. African J. Sci. 58*, 216.

Stockdale, P.H.G. 1970. Pulmonary lesions in mink with a mixed infection of *Filaroides martis* and *Perostrongylus pridhami*. *Can. J. Zool. 48*, 757–759.

Stokoe, W. J., and G.H.T. Stovin. 1944. *The caterpillars of the British butterflies, including the eggs, chrysalids and food plants.* Warne, London.

Stokoe, W. J., and G.H.T. Stovin. 1948. *The caterpillars of British moths, including the eggs, chrysalids and food plants.* vols. 1 and 2. Warne, London.

Stoltz, D. B., S. B. Vinson, and E. A. MacKinnon. 1976. Baculovirus-like particles in the reproductive tracts of female parasitoid wasps. *Can. J. Microbiol. 22*, 1013–1023.

Strong, D. R., Jr. 1974a. The insects of British trees: Community equilibration in ecological time. *Ann. Missouri Bot. Gard. 61*, 692–701.

Strong, D. R., Jr. 1974b. Nonasymptotic species richness models and the insects of British trees. *Proc. Nat. Acad. Sci. U.S.A. 71*, 2766–2769.

Strong, D. R., Jr. 1974c. Rapid asymptotic species accumulation in phytophagous insect communities: The pests of cacao. *Science 185*, 1064–1066.

Strong, D. R., Jr. 1977a. Rolled-leaf hispine beetles (Chrysomelidae) and their Zingiberales host plants in Middle America. *Biotropica 9*, 156–169.

Strong, D. R., Jr. 1977b. Insect species richness: Hispine beetles of *Heliconia latispatha*. *Ecology 58*, 573–583.

Strong, D. R., Jr., and D. A. Levin. 1975. Species richness of the parasitic fungi of British Trees. *Proc. Nat. Acad. Sci. U.S.A. 72*, 2116–2119.

Strong, D. R., Jr., and D. A. Levin. 1979. Species richness of plant parasites and growth form of their hosts. *Amer. Natur. 114*, 1–22.

Strong, D. R., Jr., E. D. McCoy, and J. R. Rey. 1977. Time and the number of herbivore species: The pests of sugarcane. *Ecology 58*, 167–175.

Struhsaker, T. T. 1967. Social structure among vervet monkeys (*Cercopithecus aethiops*). *Behaviour 29*, 83–121.

Sunderland, N. 1960. Germination of the seeds of angiospermous root parasites. In *The biology of weeds*. (Ed. by J. L. Harper), pp. 83–93. Symp. Brit. Ecol. Soc. 1. Blackwell Sci. Pub., Oxford.

Suomalainen, E. 1962. Significance of parthenogenesis in the evolution of insects. *Annu. Rev. Entomol. 7*, 349–366.

Suomalainen, E., and A. Saura. 1973. Genetic polymorphism and evolution in parthenogenetic animals. I. Polyploid Curculionidae. *Genetics 74*, 489–508.

Suomalainen, E., A. Saura, and J. Lokki. 1976. Evolution of parthenogenetic insects. *Evol. Biol. 9*, 209–257.

Tauber, C. A., and M. J. Tauber. 1977. A genetic model for sympatric speciation through habitat diversification and seasonal isolation. *Nature 268*, 702–705.

Tauber, C. A., M. J. Tauber, and J. R. Nechols. 1977. Two genes control seasonal isolation in sibling species. *Science 197*, 592–593.

Tepedino, V. J., and N. L. Stanton. 1976. Cushion plants as islands. *Oecologia 25*, 243–256.

Theodor, O., and M. Costa. 1967. A survey of the parasites of wild mammals and birds in Israel. Part I. Ectoparasites. *Israel Acad. Sci. Human. Sect. Sci.* pp. 5–117.

Thoday, J. M. 1972. Disruptive selection. *Proc. Royal Soc. London. Ser. B. 182*, 109–143.

Thomas, J. D. 1964. Studies on populations of helminth parasites in brown trout *(Salmo trutta* L.). *J. Anim. Ecol. 33*, 83–95.

Thompson, D'A.W. 1942. *On growth and form.* 2nd ed. Cambridge Univ. Press, London.

Thompson, G. B. 1938. An ectoparasite census of some common Javanese rats. *J. Anim. Ecol. 7*, 328–332.

Thompson, J. N. 1978. Within-patch structure and dynamics in *Pastinaca sativa* and resource availability to a specialized herbivore. *Ecology 59*, 443–448.

Thompson, J. N., and P. W. Price. 1977. Plant plasticity, phenology and herbivore dispersion: Wild parsnip and the parsnip webworm. *Ecology 58*, 1112–1119.

Thorpe, W. H. 1930. Biological races in insects and allied groups. *Biol. Rev. 5*, 177–212.

Tietz, H. M. 1972. *An index to the described life histories, early stages and hosts of the Macrolepidoptera of the continental United States and Canada.* vols. 1 and 2. Allyn Mus. Entomol., Sarasota, Fla.

Tomlinson, J. 1966. The advantages of hermaphroditism and parthenogenesis. *J. Theoret. Biol. 11*, 54–58.

Tostowaryk, W. 1971. Relationships between parasitism and predation of diprionid sawflies. *Ann. Entomol. Soc. Amer. 64*, 1424–1427.

Townes, H. 1969. The genera of Ichneumonidae. Part I. *Mem. Amer. Entomol. Inst. 11*, 1–300.

Townes, H., and M. Townes. 1951. Family Ichneumonidae. In *Hymenoptera of America north of Mexico.* Synoptic catalog. (Ed. by C.F.W. Muesebeck, K. V. Krombein, and H. K. Townes), pp. 184–409. U. S. Dept. Agric., Agric. Monogr. 2.

Traub, R., and C. L. Wisseman, Jr. 1974. The ecology of chigger-borne rickettsiosis (scrub typhus). *J. Med. Entomol. 11*, 237–303.

Treat, A. E. 1975. *Mites of moths and butterflies.* Cornell Univ. Press, Ithaca, N.Y.

Triantaphyllou, A. C. 1971. Genetics and cytology. In *Plant parasitic nematodes.* (Ed. by B. M. Zuckerman, W. F. Mai, and R. A. Rohde), 2:1–34. Academic Press, New York.

Uglem, G. L., and S. M. Beck. 1972. Habitat specificity and correlated aminopeptidase activity in the acanthocephalans *Neoechinorhynchus cristatus* and *N. crassus. J. Parasitol. 58*, 911–920.

Uhazy, L. S., J. C. Holmes, and J. G. Stelfox. 1973. Lungworms in the Rocky Mountain bighorn sheep of western Canada. *Can. J. Zool. 51*, 817–824.

Underwood, G. R., and F. A. Titus. 1968. Description and seasonal history of a leaf miner on poplar *Messa populifoliella* (Hymenoptera: Tenthredinidae). *Can. Entomol. 100*, 407–411.

Valentine, J. W. 1973. *Evolutionary paleoecology of the*

marine biosphere. Prentice-Hall, Englewood Cliffs, N.J.

van Someren, V.G.L. 1974. List of food plants of some East African Rhopalocera, with notes on the early stages of some Lycaenidae. *J. Lepidopt. Soc. 28,* 315–331.

van Steenis, C.G.G. 1969. Plant speciation in Malesia, with special reference to the theory of non-adaptive saltatory evolution. *Biol. J. Linn. Soc. 1,* 97–133.

Van Valen, L. 1973. A new evolutionary law. *Evol. Theory 1,* 1–30.

Vinson, S. B. 1975. Biochemical coevolution between parasitoids and their hosts. In *Evolutionary strategies of parasitic insects and mites.* (Ed. by P. W. Price), pp. 14–48. Plenum, New York.

Vinson, S. B., and J. R. Scott. 1975. Particles containing DNA associated with the oocyte of an insect parasitoid. *J. Invertebr. Pathol. 25,* 375–378.

Wade, M. J. 1977. An experimental study of group selection. *Evolution 31,* 134–153.

Wakelin, D. 1976. Host responses. In *Ecological aspects of parasitology.* (Ed. by C. R. Kennedy), pp. 115–141. North-Holland, Amsterdam.

Wallace, J. W., and R. L. Mansell (eds.). 1976. *Biochemical interaction between plants and insects.* Rec. Adv. Phytochem 10. Plenum, London.

Walsh, G. B. 1954. Plants and the beetles associated with them. In *A coleopterist's handbook.* (Ed. by G. B. Walsh and J. R. Dibb), pp. 83–98. Amat. Entomol. Soc., London.

Warner, R. E. 1968. The role of introduced diseases in the extinction of the endemic Hawaiian avifauna. *Condor 70,* 101–120.

Weigl, P. D. 1968. *The distribution of the flying squirrels Glaucomys volans and G. sabrinus: An evaluation of the competitive exclusion idea.* Ph.D. Dissertation, Duke University.

Wenzel, R. L. 1976. The streblid batflies of Venezuela (Diptera: Streblidae). *Brigham Young Univ. Sci. Bull. Biol. Ser. 20* (4), 1–177.

Wenzel, R. L., V. J. Tipton, and A. Kiewlicz. 1966. The streblid batflies of Panama (Diptera Calypterae: Streblidae). In *Ectoparasites of Panama.* (Ed. by R. L. Wenzel and V. J. Tipton), pp. 405–675. Field Museum of Natural History, Chicago.

Wertheim, G. 1970. Experimental concurrent infections with *Strongyloides ratti* and *S. venezuelensis* in laboratory rats. *Parasitology 61,* 389–395.

Whitaker, J. O., Jr. 1972. Food habits of bats from Indiana. *Can. J. Zool. 50,* 877–883.

Whitaker, J. O., Jr. 1979. Factors influencing the abundance and diversity of ectoparasites and other associates of mammals in Indiana. *Ecology.* In press.

White, M.J.D. 1954. *Animal cytology and evolution.* 2nd ed. Cambridge. Univ. Press, Cambridge.

White, M.J.D. 1968. Models of speciation. *Science 159,* 1065–1070.

White, M.J.D. 1970. Heterozygosity and genetic polymorphism in parthenogenetic animals. In *Essays in evolution and genetics in honor of Theodosius Dobzhansky.* (Ed. by M. K. Hecht and W. C. Steere), pp. 237–262. Appleton-Century-Crofts, New York.

White, M.J.D. 1973. *Animal cytology and evolution.* 3rd ed. Cambridge Univ. Press, London.

White, M.J.D. (ed.). 1974. *Genetic mechanisms of speciation in insects.* Reidel, Dordrecht.

White, M.J.D. 1978. *Modes of speciation.* Freeman, San Francisco.

Whitham, T. G. 1978. Habitat selection by *Pemphigus* aphids in response to resource limitation and competition. *Ecology 59,* 1164–1176.

Whitham, T. G. 1979. The theory of habitat selection: Examined and extended using *Pemphigus* aphids. *Amer. Natur.* In press.

Whittaker, R. H. 1969. Evolution of diversity in plant communities. *Brookhaven Symp. Biol. 22,* 178–195.

Whittaker, R. H., and P. P. Feeny. 1971. Allelochemics: Chemical interactions between species. *Science 171,* 757–770.

Wiebes, J. T. 1976. A short history of fig wasp research. *Gard. Bull. 29,* 207–232.

Wigglesworth, V. B. 1965. *The principles of insect physiology.* 6th ed. rev. Methuen, London.

Williams, C. B. 1964. *Patterns in the balance of nature and related problems in quantitative ecology.* Academic Press, London.

Williams, G. C. 1975. *Sex and evolution.* Princeton Univ. Press, Princeton, N.J.

Williams, H. H. 1960. The intestine in members of the genus *Raja* and host-specificity in the Tetraphyllidea. *Nature 188,* 514–516.

Williams, H. H., A. H. McVicar, and R. Ralph. 1970. The alimentary canal of fish as an environment for helminth parasites. In *Aspects of fish parasitology.* (Ed. by A.E.R. Taylor and R. Muller), pp. 43–77. Symp. Brit. Soc. Parasitol. 8.

Williams, N. D., F. J. Gough, and M. R. Rondon. 1966. Interaction of pathogenicity genes in *Puccinia graminis* f. sp. *tritici* and reaction genes in *Triticum aestivum* ssp. *vulgare* "Marquis" and "Reliance." *Crop Sci. 6,* 245–248.

Wilson, A. C., G. L. Bush, S. M. Case, and M.-C. King. 1975. Social structuring of mammalian populations and rate of chromosomal evolution. *Proc. Natl. Acad. Sci. U.S.A. 72,* 5061–5065.

Wilson, E. O. 1969. The species equilibrium. *Brookhaven Symp. Biol. 22,* 38–47.

Wilson, J. W. 1951. *Micro-organisms in the rhizosphere of beech.* Ph.D. Thesis, Oxford University.

Wood, R.K.S., and A. Graniti (eds.). 1976. *Specificity in plant diseases.* Plenum, London.

Wright, C. A. 1971. *Flukes and snails.* Allen and Unwin, London.

Wright, C. A., and V. R. Southgate. 1976. Hybridization of schistosomes and some of its implications. In *Genetic aspects of host-parasite relationships.* (Ed. by A.E.R. Taylor and R. Muller), pp. 55–86. Symp. Brit. Soc. Parasitol. 14.

Wright, S. 1931. Evolution in Mendelian populations. *Genetics 16,* 97–159.

Wright, S. 1940. Breeding structure of populations in relation to speciation. *Amer. Natur. 74,* 232–248.

Wright, S. 1943. Isolation by distance. *Genetics 28,* 114–138.

Wright, S. 1949. Adaptation and selection. In *Genetics, paleontology and evolution.* (Ed. by G. L. Jepson, E. Mayr, and G. G. Simpson), pp. 365–389. Princeton Univ. Press, Princeton, N.J.

Wylie, H. G. 1960. Insect parasites of the winter moth, *Operophtera brumata* (L.) (Lepidoptera: Geometridae) in western Europe. *Entomophaga 5,* 111–129.

Yeo, P. F. 1978. A taxonomic revision of *Euphrasia* in Europe. *Bot. J. Linn. Soc. 77,* 223–334.

Zeigler, B. P. 1977. Persistence and patchiness of predator-prey systems induced by discrete event population exchange mechanisms. *J. Theoret. Biol. 67*, 687–713.

Zimmerman, E. C. 1938. Cryptorhynchinae of Rapa. *Bull. Bernice P. Bishop Mus. 151*, 1–75.

Zimmerman, E. C. 1960. Possible evidence of rapid evolution in Hawaiian moths. *Evolution 14*, 137–138.

Zinsser, H. 1935. *Rats, lice and history*. Little, Brown, Boston.

Zwölfer, H. 1975. Speciation and niche diversification in phytophagous insects. *Verh. Deutsch. Zool. Ges. 67*, 394–401.

Zumpt, F. 1965. *Myiasis in man and animals in the Old World. A textbook for physicians, veterinarians, and zoologists*. Butterworths, London.

Author Index

Alexopoulos, C. J., 60
Alstad, D. N., 51
Anderson, R. C., 154, 155
Anderson, R. M., 9, 46-48, 74, 150
Andrewartha, H. G., 9, 66
Arentzen, J. C., 173
Arme, C., 71
Arthur, D. R., 5
Ashton, D. H., 169
Askew, R. R., 5, 8, 18, 25, 28, 29, 87, 112, 120, 160, 162-64
Atchley, W. R., 85
Atsatt, P. R., 34, 57, 87, 150

Baer, J. G., 80
Baker, H. G., 22
Barbehenn, K. R., 157-59, 162, 166
Barnes, H. F., 17, 29
Barnes, M. M., 101
Batra, L. R., 49, 150
Batra, S.W.T., 49, 150
Beaver, R. A., 25
Beck, S. M., 140
Beddington, J. R., 44, 45, 52
Bellotti, A. C., 36
Benson, R. B., 94
Benson, W. W., 27, 131
Bequaert, J. C., 21, 30, 120, 122
Berlocher, S. H., 102
Berrie, A. D., 55, 56, 74, 76
Bethel, W. M., 150
Bindernagel, J. A., 154
Birch, L. C., 9, 66
Björkman, E., 150
Blackman, R. L., 102
Bonner, J. T., v, vii, 85
Bouček, Z., 112
Bretsky, P. W., 11
Brinck, P., 33
Brinkerhoff, L. A., 101
Brittain, E. G., 165-67
Broekhuizen, S., 158

Brown, K. S., 27, 131
Brown, W. J., 25
Brues, C. T., 34
Buchner, P., 49, 132, 133, 173
Buechner, H. K., 153
Burdon, J. J., 157, 162, 165-68
Burnet, M., 170
Burns, D. P., 29
Burt, W. H., 154
Bush, G. L., 25, 26, 40, 41, 55, 97-101
Busvine, J. R., 170

Calow, P., 16
Campbell, E. O., 150
Carson, H. L., 25, 38, 95
Caswell, H., 46
Cates, R. G., 34, 51
Chaplin, S. J., 77
Chappell, L. H., 74, 137, 139, 143, 144
Charnov, E. L., 46
Chilvers, G. A., 157, 162, 165-68
Chubb, J. C., 137, 142
Cifelli, R., 11
Cisne, J. L., 11, 12
Clapham, A. R., 35
Claridge, M. F., 33, 109
Clarke, B., 20
Clausen, C. P., 17
Clay, T., 33, 119
Cleveland, L. R., 42
Cody, M. L., 148, 159
Cohen, J. E., 50, 74, 75, 159
Cohen, S. H., 94
Coles, J. F., 51
Colinvaux, P. A., 9
Collins, N. C., 51, 52, 59, 150
Cooper, A. F., 17
Cope, E. D., 11, 12
Corbet, A. S., 131
Cornell, H. V., 109, 159, 162
Costa, M., 21, 30, 119, 120

227

Cowdry, E. V., 132
Craddock, E. M., 102, 103
Craigie, J. H., 62
Cram, E. B., 21
Crofton, H. D., 20, 47, 74
Crompton, D.W.T., 139
Crosby, A. W., 152, 158, 170
Cross, S. X., 136, 143
Crozier, R. H., 13
Cuellar, O., 84
Culver, D., 46

Darlington, A., 29
Darwin, C., 15, 28, 128, 159
Das, K. M., 20
Davis, D. M., 79
Day, P. R., 5, 34, 38, 42, 62, 92
DeVay, J. E., 100
Deverall, B. J., 34, 62, 100, 149
Dobzhansky, T., 91, 92, 130, 159
Dogiel, V. A., 30, 34, 70, 173
Dowell, R. V., 117
Dritschilo, W., 32, 108, 109
Duke, B.O.L., 98, 101
Dupuis, C., 29
Dybas, H. S., 17

Eastop, V. F., 37
Edmunds, G. F., 51
Ehrlich, P. R., 34, 70
Eibl-Eibesfeldt, E., 150
Eibl-Eibesfeldt, I., 150
Eichler, W., 28, 37
El Mofty, M., 42
Elton, C. S., 9, 152
Emlen, J. M., 46
Endler, J. A., 40
Euzet, L., 136
Evans, F. C., 70
Evans, G. C., 169
Evans, J. W., 101

Fabre, J. P., 51
Fagg, P. C., 170
Feeny, P., 34, 51, 149
Fenner, F., 10, 55, 150
Fischer, M., 112
Flessa, K. W., 11, 12
Flor, H. H., 34, 38
Ford, B., 5, 42

Ford, J., 154
Ford, L. T., 21, 29-31
Forrester, D. J., 152
Foster, A. O., 136
Foster, M. S., 30, 42
Fowler, D. P., 51
Frandsen, F., 56
Frankland, H.M.T., 137, 143
Franklin, M. T., 80, 81, 94
Free, C. A., 44, 45
Freeland, W. J., 150
Frings, H., 173
Futuyma, D., 51

Galun, R., 19
Garrett, S. D., 6, 17
Ghiselin, M. T., 18
Gibson, C.W.D., 70
Gibson, D. I., 74, 135, 137, 147
Gibson, L. P., 29
Giesbert, E. F., 97
Gilbert, L. E., 27, 79, 98, 131, 160, 162
Gill, D. E., 158
Glazier, R. A., 20
Glesener, R. R., 84, 94
Goldberg, E., 173
Goodpasture, C., 102
Gordh, G., 25
Gough, F. J., 62
Gould, S. J., 14, 87, 88
Govier, R. N., 119
Graham, K., 49, 150
Graham, M.W.R.deV., 112
Graniti, A., 119
Grant, V., 40
Green, G. J., 62, 100
Griffiths, G.C.D., 21, 112, 127-29
Grissell, E. E., 102
Grossenheider, R. P., 154
Grubb, P. J., 51
Guevrement, M., 117
Gurney, W.S.C., 46

Hair, J. D., 74, 136, 141, 142
Hairston, N. G., 74
Halvörsen, O., 33
Hamilton, W. D., 19
Hammond, P. S., 52
Harley, J. L., 6

Harper, J. L., 62, 69, 76, 86
Harrison, L., 33
Hassell, M. P., 16
Hastings, A., 46
Heatwole, H., 79
Hegnauer, R., 35
Heinrich, G. H., 21
Helle, W., 76, 96, 97
Hering, E. M., 30, 31, 35, 36
Hershkovitz, P., 53
Heywood, V. H., 35, 110
Hibler, C. P., 152
Hildahl, V., 29
Hincks, W. D., 6, 7, 19
Hirst, J. M., 69
Hobbs, R. P., 74
Holmes, J. C., 19, 72, 74, 134, 136-41, 143, 144, 150, 152
Hoogstraal, H., 5
Hopkins, G.H.E., 32, 33, 119-21
Horn, H. S., 46
Hovore, F. T., 97
Huettel, M. D., 100
Huffaker, C. B., 11, 16, 47
Hutchinson, G. E., 15, 46, 105, 134, 141, 146
Huxley, J. S., 12, 13

Istock, C. A., 147
Iwata, K., 118

Jaenike, J., 93, 94, 162
Janzen, D. H., 28, 32, 113, 118, 126, 127, 130, 161, 162
Jenkins, D., 158
Jennings, J. B., 16
Joern, A., 161
John, B., 84
Johnston, D. E., 150
Jones, A. W., 17, 25
Jordan, K., 33

Kagan, I. G., 57
Kao, K. N., 62
Kassaby, F. Y., 170
Kellog, V. L., 32
Kemmers, R., 158
Kennedy, C. R., 5, 9, 17, 21, 47, 70, 71, 74, 175
Kethley, J. B., 150

Kiewlicz, A., 120
Kinsey, A. C., 40
Kloet, G. S., 6, 7, 19
Klopfer, P. H., 130
Knott, D. R., 62
Koford, C. B., 162
Kostrowicki, A. S., 70
Krebs, C. J., 9
Kuijt, J., 5
Kuntz, R. E., 141

LaMarche, V. C., 169
Lange, R. E., 152
Lankester, M. W., 154
Large, E. C., 62
Lawton, J. H., 30, 32, 33, 44, 45, 52, 69, 70, 108, 109, 111, 112, 160, 161
Lees, A. D., 42
LeJambre, L. F., 20
Lejeune, R. R., 29
Leong, T. S., 74
Leston, D., 21, 29
Levin, D. A., 30, 32, 94, 109
Levin, S. A., 48
Levins, R., 19, 20, 46
Lewis, T., 29, 78, 79, 90
Limbaugh, C., 150
Lin, N., 150
Lindquist, E. E., 17, 33, 78, 150
Linsley, E. G., 25, 33
Llewellyn, J., 140
Lloyd, M., 17
Loegering, W. Q., 62
Lokki, J., 13, 85
Lorenz, D. M., 11
Luig, N. H., 62

MacArthur, R. H., v, 19, 22, 24, 32, 44, 46, 65, 106, 130, 141, 159
Macauley, B. J., 169
MacInnis, A. J., 19
MacKenzie, K., 74, 135, 137, 147
Mackie, R. A., 151
MacKinnon, E. A., 50, 132
Maddison, S. E., 57
Malavasi, A., 42
Mansell, R. L., 34
Manter, H. W., 33

Marchalonis, J. J., 149
Marks, G. C., 170
Marshall, A. G., 119
Martens, J. W., 62
Martin, A. C., 21
Martin, D. R., 136
May, R. M., 4, 8, 16, 44, 47, 52
Maynard Smith, J., 40, 82, 85, 93, 94
Mayr, E., v, 11, 25, 33, 37, 39-41, 76, 77, 84, 95, 103
Mayse, M. A., 69
McClure, M. S., 70
McCoy, E. D., 32, 107, 109
McKenzie, R.I.H., 62
McLeod, J. M., 66, 67
McNaughton, S. J., 70
McNeill, W. H., 170
McVicar, A. H., 74, 135, 137
Melander, L. W., 62
Mellinger, M. V., 70
Metcalf, M. M., 33
Metzger, C. J., 152
Michener, C. D., 150
Miller, G. R., 158
Mitchell, R., 17, 51, 52, 59, 78
Mitter, C., 51
Moericke, V., 79
Mook, J. H., 52
Moore, W. C., 6
Morris, J. R., 71
Morris, M. G., 29
Morris, R. F., 88, 150
Morrow, P. A., 168, 169
Mulligan, H. W., 154
Murdoch, W. W., 70
Murray, J. S., 29
Murray, M. D., 150

Naumov, N. P., 9
Nechols, J. R., 42
Nei, M., 96
Nelson, A. L., 21
Nelson, B. C., 150, 158
Newell, N. D., 11, 12
Nisbet, R. M., 46
Noble, E. R., 9, 10, 28, 33, 37, 80, 143
Noble, G. A., 9, 10, 28, 34, 37, 80, 143

Nowakowski, J. T., 114

Odum, E. P., 9
Oliver, J. H., 13, 18
Opler, P. A., 32, 108, 109, 114
Otte, D., 161
Owen, D. F., 126, 173
Owen, J., 126
Owen, R. W., 71

Pamilo, P., 13
Paperna, I., 137, 143-46
Park, T., 156
Parker, E. D., 94
Parks, D. R., 70
Pearsall, W. H., 153
Pendlebury, H. M., 131
Perring, F. H., 108
Peterson, C. H., 70
Petter, A. J., 75, 136, 141
Phillips, P. A., 101
Pianka, E. R., v, 46, 159
Pickett, S.T.A., 49, 51
Pielou, E. C., 141
Pieterse, A. H., 76, 96, 97
Pimentel, D., 36
Pimm, S. L., 52
Pond, C. M., 126
Poole, R. W., 141
Powell, J. A., 151
Powell, J. R., 25, 94
Powell, N. T., 50
Powers, H. R., 62
Powers, K. V., 11, 12
Prestwood, A. K., 154, 155
Price, P. W., vi, 7, 16, 19, 27, 31, 33, 36, 39, 52, 58, 66, 68-70, 85, 97, 108, 109, 111, 112, 115, 117, 118, 124, 127
Prokopy, R. J., 79
Prop, N., 150
Pryor, L. D., 165
Pulliam, H. R., 46
Pyke, G. H., 46

Quednau, F. W., 117

Ralph, R., 74, 135, 137
Ratcliffe, F. N., 10, 150
Ratcliffe, L. H., 20

Rathcke, B. J., 52, 70, 127
Raup, D., 12
Raven, P. H., 34
Rey, J. R., 32, 107, 109
Rhoades, D. F., 34, 51
Richards, C. S., 56
Richards, P. W., 172
Riom, J., 51
Rishbeth, J., 17
Roche, L., 51
Rohde, K., 147, 148
Rondon, M. R., 62
Root, R. B., 77
Rosenfield, P. L., 46
Rosengren, R., 13
Ross, H. H., 25, 37, 40, 101
Roth, L. M., 94
Rothschild, M., 42, 120, 122
Royama, T., 46
Rumpus, A., 47
Ruse, J. M., 29

Salt, G., 9, 42
Sankurathri, C. S., 72
Saura, A., 13, 85
Saville, D.B.O., 33
Schad, G. A., 75, 135, 136, 140, 141
Schaller, G. B., 150
Schnathorst, W. C., 100
Schneider, J. C., 51
Schoener, T. W., 134, 148
Schopf, T.J.M., 12
Schröder, D., 30, 32, 109, 161
Scott, G. R., 153
Scott, J. R., 50
Scriber, J. M., 130
Scudder, G.G.E., 29, 40
Selander, R. K., 93
Shapiro, A. M., 150
Sharp, M. A., 70
Shattock, R. C., 92
Shields, O., 79, 80
Simberloff, D. S., 24
Simmonds, H. W., 170
Simpson, G. G., 11, 12
Singer, M. C., 98
Skellam, J. G., 46, 146
Slansky, F., 70
Slatkin, N. G., 46

Smith, C. R., 158
Smith, N. G., 151, 173
Smith, R. A., 46
Smithers, S. R., 57
Smyth, J. D., 42
Snyder, T. P., 13
Sogandares-Bernal, F., 74, 136
Sonneborn, T. M., 85
Southgate, V. R., 56, 57, 98
Southwood, T.R.E., 21, 29, 30, 33
Spencer, K. A., 21, 108, 128, 129
Spieth, H. T., 40
Spinage, C. A., 153
Stanton, N. L., 109
Stebbins, G. L., 22, 40, 43, 86
Steinhaus, E. A., 132, 133
Stelfox, J. G., 152, 153
Stern, K., 51
Stevenson-Hamilton, J., 154
Stockdale, P.H.G., 136, 139
Stokoe, W. J., 29, 131
Stoltz, D. B., 50, 132
Stovin, G.H.T., 29, 131
Strong, D. R., 27, 30, 32, 33, 105, 107, 109
Struhsaker, T. T., 150
Sunderland, N., 17
Suomalainen, E., 13, 18, 85

Tauber, C. A., 41, 42
Tauber, M. J., 41, 42
Tepedino, V. J., 109
Theodore, O., 21, 30, 119, 120
Thoday, J. M., 40
Thomas, J. D., 137, 142
Thompson, D'A. W., 3
Thompson, G. B., 33
Thompson, J. N., 49, 52, 58
Thorpe, W. H., 25
Tietz, H. M., 131
Tilman, D., 84, 94
Tipton, V. J., 120
Titus, F. A., 29
Tomlinson, J., 82
Tostowaryk, W., 150
Townes, H., v, 8, 21, 28
Townes, M., 21
Traub, R., 23, 53
Treat, A. E., 27, 79

Triantaphyllou, A. C., 80, 81, 94, 102
Tutin, T. G., 35

Uglem, G. L., 140
Uhazy, L. S., 152
Underwood, G. R., 29
Usinger, R. L., 25, 33

Valentine, J. W., 11
Van Gundy, S. D., 17
van Someren, V.G.L., 131
van Steenis, C.G.G., 151
Van Valen, L., 94
Vespäläinen, K., 13
Vinson, S. B., 19, 50, 132

Wade, M. J., 96
Wakelin, D., 34
Wallace, J. W., 34
Walsh, G. B., 29
Walters, S. M., 108
Warburg, E. F., 35
Warner, R. E., 159
Washburn, J. O., 109
Watson, A., 158
Watson, I. A., 62
Weigl, P. D., 173
Wells, W. H., 141
Wenner, A. M., 79
Wenzel, R. L., 120
Wertheim, G., 136
Whitaker, J. O., 21, 30, 39
White, D. O., 55, 170
White, M.J.D., 14, 18, 26, 39, 40,

41, 76, 82, 84, 85, 95, 101, 103, 172
Whitham, T. G., 173
Whitlock, J. H., 20
Whittaker, R. H., 34, 149, 163
Wiebes, J. T., 151
Wiegert, R. G., 51, 52, 59, 173
Wigglesworth, V. B., 42
Williams, C. B., 130
Williams, G. C., 16, 18, 82, 83, 84
Williams, H. H., 74, 135-37, 140
Williams, N. D., 62
Willson, M. F., 97
Wilson, A. C., 97
Wilson, E. O., v, 19, 22, 24, 32, 44, 65, 106, 138
Wilson, J. W., 70
Wilson, M. R., 33, 109
Wisseman, C. L., 23, 53
Wistrand, H., 94
Wolman, M. G., 46
Wood, R.K.S., 119
Worms, M. J., 57
Wright, C. A., 56, 57, 88, 98
Wright, S., 25, 96
Wylie, H. G., 28, 29

Yeo, P. F., 57, 87

Zeigler, B. P., 46
Zim, H. S., 21
Zimmerman, E. C., 25, 101
Zinsser, H., 170
Zwölfer, H., 161
Zumpt, F., 120, 121

Subject Index

Adaptive radiation, 3, 14, 26-39, 41, 43, 105-33; and mutualism, 132, 133; in tropics, 126
Adelges, 22
Adelina tribolii, 156, 157
age specificity of parasites, 52, 70
allochronic isolation, 41, 100
Alsophila pometaria, 51
Alternaria tenuis, 50
Amphimixis, 80, 81
Apocreadium, 136, 140
Aporocotyle, 136, 140
Armillaria mellea, 6
Ascaris lumbricoides, 91
Atractis dactyluris, 141

Berberis, 60, 61, 63
bighorn sheep and disease, 152, 153
biogeographic effects of parasites, 153-59, 173
biological progress, 12-14
Biomphalaria, 56
biotic associations of parasites, 49, 50; complexity of, 48-53, 58, 61, 72, 132
broken stick model, 141
browsers, 4, 45
Brugia, 22
Bulinus, 76

Calycotyle, 136
Capillaria, 137
Castroia, 136
Cediopsylla simplex, 42
Cephenemyia trompe, 121
Cerodontha pygmaea, 129
cervids and disease, 154-56
Ceuthospora innumera, 169
Chaetogaster limnaei, 73
chromosomal rearrangements, 96, 97, 102, 103

coevolution, 9, 10, 26, 34-39, 50, 138
coexistence, 20, 32, 46, 65, 106, 113, 134-48; non-equilibrium, 146, 147; of *Eucalyptus* species, 168, 169
colonization probability, 22, 24, 30, 32, 48, 63-66, 70-78, 113, 131, 152; of specialized hosts, 127-29
community development phases, 138
community maturity, 134
community organization, 134-48; interactive, 136-38, 143-47; noninteractive, 134-43
competition, 9, 10, 32, 67, 71, 74, 94, 110, 113, 134, 138
competitive exclusion, 137, 143-47
complex life cycles, 4, 22
convergent evolution, 13
cost of evolution, 88, 89
Crenosoma hermani, 139
Crepidostomum, 137
Cronartium ribicola, 22
Cucullanus, 137, 147
Culex pipiens, 159
Cylicocyclus, 136
Cynips, 160, 163-65

Dactylogyrus, 137, 144, 145
Depressaria pastinacella, 57
Dicrocheles phalaenodectes, 78
Didymuria violescens, 102, 103
Diplostomum spathaceum, 17
diseases of large animals, 151-56
dispersal, 16, 17
dispersion of parasites, 47
disruptive selection, 40, 41
distributional gaps between hosts, 159, 160
diversity of parasites, 16, 28, 39, 70, 80, 126

233

diversity of species, 15, 159

Echinoparyphium recurvatum, 71
Echinorhynchus, 137, 142
ecological isolation, 57, 97, 101, 155
ecological niche of parasites, 134-48; fundamental and realized, 134, 135, 138, 143, 144, 159. *See also* niche
Eichler's rule, 28
environmental grain, 10, 20
environmental predictability, 23. *See also* resources
ephemeral patches, 48, 52-54, 58, 59, 66, 77, 78, 84
epidemiology, 152; and distance between hosts, 162; of *Neodiprion swainei*, 66-69; of scrub typhus (Rickettsiosis), 23, 48, 53-55; of viruses, 55
equilibrium, 15, 20, 24, 44-47, 74, 75, 106, 110, 114; definitions, 44, 45, 65
Eriosoma, 22
Erysiphe graminis, 157
Erythroneura, 25
Euceraphis, 102
Eulia fratria, 57
Euphrasia, 87
evolutionary rate, 12, 13, 24-26, 36, 37, 41, 62, 77, 89, 96, 104, 150; reduced in parasites, 86; slow in host, 85
evolutionary time, 26, 32-34
extinction probability, 24, 48, 65, 70, 77, 95, 96
extinction rate, 11, 44, 47, 48, 95, 96, 111

Fahrenholz's rule, 33, 34
fecundity, 24, 86, 87, 90
Filaroides, 136, 139
food chain length, 52
founder effect, 95
Fuhrman's rule, 34
Fusarium oxysporum, 17
Fusarium wilt, 50

gall wasps and parasites, 160-65
gene fixation, 95, 96
gene flow, 17, 77, 95, 96, 104
gene-for-gene concept, 38, 42, 62, 86, 101
general concepts on parasites, 15-43; ecological, 16-24; evolutionary, 24-43
generalization, 28, 34, 37
generation time, 10, 24, 86, 90
genetic distance between hosts, 162
genetic drift, 95
genetic systems, 9, 13, 19, 57, 76-104; comparison between parasites and predators, 104; definition, 77
genetic variation; and mutation rates, 91; in haplodiploid species, 93; in host, 175; in parthenogenetic species, 85; in parasite progeny, 81-87, 89, 91, 92, 94
geographic races, 20, 24
Graphidium strigosum, 158
grazers, 4, 20, 21
group (= interdemic) selection, 96, 158, 173

habitat heterogeneity, 40; complexity, 65
Haemonchus contortus, 20
haplodiploid species, 93
Hedylepta, 25
Heliconius, 27, 79, 130, 131
Helisoma trivolvis, 71
Heterodera, 80
heteroecious life cycles, 49
heteroecism, 59, 60
host, age, 52, 70; as environment for parasite, 9; defense, 149; diets and parasites, 173; diversity, 26-28, 118, 175; effects on parasite, 97, 98; mobility, 39; -parasite models, 47, 48, 165-68; phenology, 41; predisposed to parasites, 50, 170; races, 25, 33, 40-42, 56, 57, 62; resistance, 36, 56, 62, 63; resistance to polymorphism,

56; selection genes, 98, 99; shifts, 40, 41, 98-101, 173; survival genes, 98, 99; susceptibility differential, 153-58, 166, 167; susceptibility differential to predation, 150, 154, 158; target size, 26, 29-34; target body size, 26, 29-30, 32, 33; target geographical range, 26, 30-32, 106-13; target population size, 26, 30-32, 106; topography, 12
Hymenolepis, 136, 137, 142, 144
Hyphantria cunea, 88-89

immune response, 9, 38, 57, 139, 144, 146, 149
inbreeding, 13, 18, 24, 51, 79, 84
index of specificity, 123
interactive communities, 136-38, 143-47
interdemic selection, 96, 158, 173
intrinsic rate of natural increase, 88
Itygonimus, 137

Kalicephalus, 136, 140

Laspeyresia pomonella, 101
latitudinal trends; diversity of parasites, 126, 127; species diversity, 147; specificity, 130, 131
law of the unspecialized, 11
Leishmania, 22, 25, 26
leishmaniasis, 25
Lepidapedon, 137
Leptotrombidium deliense, 23
life cycle; complex, 4, 22; heterogonic, 18
Ligula intestinalis, 71
Limothrips denticornis, 79
Loa loa, 98, 101
Longevity; of parthenogenetic species, 93; of resting stages, 17
long-term studies, 70, 175
lungworm-pneumonia complex, 152

Massospora cicadina, 17

mating, before dispersal, 78, 79; at recognizable sites, 79, 80
Megarhyssa, 79
Mehdiella uncinata, 141
Meloidogyne, 80, 94
meningeal worm, 154-56
mobility of parasites, 19, 119, 121, 122
Moniliformes, 137, 144
mutation rate, 91
mutualism, 6, 49, 50, 72, 73, 132, 133, 150, 151; and adaptive radiation, 132, 133
mycetome, 132, 133
mycorrhizal fungi, 6, 150
myxoma virus, 10, 150

Nematus, 94
Neodiprion swainei, 65
Neoechinorhynchus, 136, 137
niche (*see also* ecological niche); breadth, 10; overlap, 135-37, 140-44; segregation, 134, 135, 138-48
non-equilibrium, 44-75; communities, 9, 22, 64-73, 106, 113; definition, 44; models, 46; populations, 9, 22, 24, 44, 64, 86, 94
noninteractive communities, 134-43
Nuculaspis californica, 51

Oedomagena tarandi, 121
Oestrus aureoargentatus, 121
Omphalometra, 137
Onchocerca, 22
Operophtera brumata, 28
optimal searching strategies, 46
Orthocarpus, 87
outbreeding, 79, 87, 85

Paracoenia turbida, 59
parasite, definition, 4, 5; differential pathogenicity, 153-58, 166, 167; effects on host competition, 156-58, 165-67, 170; food chain trends, 43; impact on host, 149-70
parasite-free space, 161

parasite-host models, 47, 48, 165-68

parasites, as agents in biological warfare, 172, 173; as predators, 9; graphical model of distance between hosts, 161; maintain distance between hosts, 159-69; of *Eucalyptus*, 168-70; percent in fauna, 7, 8, 80; select for parthenogenesis, 172

parasitic angiosperms, 5, 17, 34, 57, 150

parasitic castration, 71

parasitological rules; Eichler's, 28; Fahrenholz's, 33, 34; Furhman's, 34; Szidat's, 37

Parelaphostrongylus tenuis, 154-56

parthenogenesis (*see also* reproduction); adaptive nature of, 82-95; commonness of, 93, 94; criticism, 93-94; long term, 85-88; short term, 82-85

Partnuniella thermalis, 59

patch dynamics, 16, 45, 47, 49, 54-62

patchy resources, 48, 50-54, 58, 59, 77, 78, 84, 113

Pemphigus, 22

percent specificity, 123

Perostrongylus pridhami, 139

Pharyngobolus africanus, 121

phoresy, 17, 78

Phyllonorycter, 28

phylogeny of hosts, 33

Phymatotrichum omnivorum, 17

Physa gyrina, 73

Phytomyza ranunculi, 129

Phytophthora cinnamomi, 169

Phytophthora infestans, 92

Piggotia substellata, 169

Pineus, 22

Plasmodium, 22

polymorphism, 20, 51, 77

polyploid species, 13

Pomphorynchus laevis, 46

population, definition, 76; effective size, 96; growth, 88; size, 10, 17, 44, 74, 76, 77, 79, 95; structure, 77, 95, 174, 175

Pratylenchus, 94

predators, 4, 7, 9, 18-21, 26, 37, 45, 94; compared to parasites, 77, 104

progenesis, 87, 88

Proteocephalus, 136, 137

Protostrongylus stilesi, 152

Puccinia graminis, 22, 60-64, 100

Puccinia hordei, 22, 92

Pycnoscelus, 94

rape, 79

rare events, 70, 171, 174, 175

recombinational load, 83, 84, 93

reproduction, types of, in parasites, 17, 18, 77, 78, 81, 82, 104; amphimixis, 80, 81; automixis, 81; arrenotoky, 18, 24, 93; asexual, 13, 18, 24; inbreeding, 13, 18, 24, 51, 79, 84; cyclical parthenogenesis, 18, 81, 82; definitions, 81; hermaphroditism, 78, 80; in nematodes, 80, 81; parthenogenesis, 13, 18, 19, 51, 78, 80, 82-95; polyembryony, 18; pseudogamy, 82; thelytoky, 81, 87, 90, 91, 95; triggered by host, 42

reproductive isolation, 41, 98, 100, 103; potential, 10, 90; rate, 24, 82

resource partitioning, 134, 135, 138-48

resources, ephemeral, 22, 48, 52-54, 58, 59, 66, 77, 78, 84; heterogeneity of, 48; patchy, 19, 48, 50-54, 58, 59, 77, 78, 84, 113; predictable, 83, 84; rarity of, 126, 127, 131; unpredictable, 43, 83; unsaturated, 65, 69-71, 94, 105, 106, 113, 115-18, 138, 140, 147, 148

Rhabdias bufonis, 80

Rhagoletis indifferens, 25

Rhagoletis pomonella, 98, 99, 102

Rhinoestrus giraffae, 121

Rhinoestrus hippopotami, 121

Rhizoctonia solani, 6

Rickettsia tsutsugamushi, 23, 53, 54
rickettsiosis, 23, 48, 53-55
rinderpest, 153, 154; effect on trypanosomiasis, 154

saprophages, 4, 7
Schistosoma, 55-57, 60, 74; *haematobium*, 46, 57, 76, 98; hybridization, 56, 57; *intercalatum*, 57, 98; *mansoni*, 56; patch dynamics, 55-57; *rodhaini*, 56
schistosomiasis, 75
sex ratio, 78, 79, 87, 90, 92
Sitodiplosis mosellana, 17
size and number relationship, 8
specialization, 10, 11, 19, 20, 26, 32-39, 51, 57, 118-33; and life history traits, 119-22
speciation, 33, 39-43, 78; allopatric, 39, 40, 41, 101; chromosomal rearrangements, 102, 104; geographic, 39, 40, 41, 101; index, 124, 125; parapatric, 55, 103, 104; rates, 24-26, 34; sympatric, 39-43, 98-104; preadaptive traits for sympatric, 101
Species, packing, 10, 20, 32, 74, 75, 105-107, 134-48; sibling, 25, 33, 41; subspecies, 25
species-area relationship between parasites and hosts, 106-14
specificity, 105-33, 168; host age, 52; index of, 123; latitudinal trends in, 130, 131; on two trophic levels, 126-32; percent, 123
Spilopsyllus cuniculi, 42
Splendidofilaria, 22
stability, 15, 52, 88; population, 10, 23, 44, 46; community, 10
Strongyloides, 80, 136
survivorship curves, 66, 115, 116, 118
Szidat's rule, 37

Tachygonetria, 135, 136, 141
Tetraopes linsleyi, 97
Tetraopes tetrophthalmus, 97
Thrips tabaci, 90-93
Trypanosoma, 22

Wuchereria, 22

Xanthomonas malvacearum, 100

Yersinia pestis, 158

Library of Congress Cataloging in Publication Data

Price, Peter W
    Evolutionary biology of parasites.

    (Monographs in population biology; 15)
    Bibliography: p.
    Includes indexes.
    1. Parasites—Evolution.   I. Title.
II. Series.
QL757.P74   574.5'24   79-3227
ISBN 0-691-08256-1
ISBN 0-691-08257-X pbk.